BOUNDARY ELEMENT METHODS IN CREEP AND FRACTURE

Consulting Editor
J. H. GITTUS
UKAEA, Springfields, Preston, UK
Editor-in-Chief
Res Mechanica: The International Journal of Structural Mechanics and Materials Science

BOUNDARY ELEMENT METHODS IN CREEP AND FRACTURE

SUBRATA MUKHERJEE

Department of Theoretical and Applied Mechanics,
Thurston Hall,
Cornell University,
Ithaca, NY 14853, USA

APPLIED SCIENCE PUBLISHERS
LONDON and NEW YORK

APPLIED SCIENCE PUBLISHERS LTD
Ripple Road, Barking, Essex, England

Sole Distributor in the USA and Canada
ELSEVIER SCIENCE PUBLISHING CO., INC
52, Vanderbilt Avenue, New York, NY 10017, USA

British Library Cataloguing in Publication Data

Mukherjee, Subrata
Boundary element methods in creep and fracture.
1. Engineering mathematics
2. Boundary value problems
I. Title
620′.001′515353 TA347.B69

ISBN 0-85334-163-X

WITH 24 TABLES AND 73 ILLUSTRATIONS

Photoset by Interprint Limited, Malta
Printed in Great Britain by Galliard (Printers) Ltd, Great Yarmouth

To Krishna, who makes the wheels turn

Preface

This book is concerned with the applications of the boundary element method to problems of time-dependent inelastic deformation and fracture of metallic media. Such problems are of interest in many fields including energy, turbomachinery and aerospace.

Although the boundary element method is rooted in classical integral equation formulation of problems, numerical implementation of the method is of fairly recent vintage. Roughly speaking, the method has been used to obtain numerical solutions to linear problems in solid mechanics in the 1960s and problems with material nonlinearities in the 1970s. I and my coworkers have been intimately involved with applications of the method to solve nonlinear problems in solid mechanics, particularly in viscoplasticity and fracture, for the last seven years.

This book is divided into ten chapters. The first four chapters set the stage for applications to specific classes of problems. These are followed by four chapters dealing with the applications to planar, axisymmetric, torsion and plate bending problems. The last two chapters are concerned with inelastic fracture mechanics.

The strength or weakness of a numerical method can only be judged through obtaining numerical solutions to specific problems. With this in view, many numerical applications of the method are presented in the book. The results obtained by the boundary element method are compared to those obtained by other methods whenever possible. In particular, the more widely used finite element method is given special attention and comparisons of results from the boundary element and finite element methods, with regard to accuracy and computational efficiency, have been carried out in many cases. Both these are powerful general-purpose methods and, as can be expected, the boundary element method appears superior in some cases while the finite element method performs better in others. In many applications, both methods appear to be equally efficient.

This book is directed towards researchers, practising engineers and scientists, and postgraduates. The reader is expected to be familiar with the general area of solid mechanics and with the basic techniques of applied mathematics and numerical methods.

I wish to thank a number of people and organizations who have contributed in various ways to make this book possible. First, my principal research collaborators, my former PhD students Dr V. Kumar and Dr M. Morjaria, who have made significant contributions to much of our group's research that has been described in this book. Next, my current students V. Sarihan and V. Banthia, for their contributions to the work on axisymmetric deformation and inelastic fracture, respectively, and A. Chandra and S. Ghosh for their help during the writing of this book. I am indebted to my colleagues at Cornell, Professors Hart, Li, Conway, Moon and Pao, for their encouragement, and to Professors Banerjee and Shaw of Buffalo and Rizzo and Shippy of Kentucky for several useful technical discussions. I greatly appreciate the terrific artwork of Mrs Jane Jorgensen who was never a day late, and the excellent typing of Ms Delores Hart who typed so many long equations accurately and cheerfully. Finally, I appreciate the effort of Ms Hillary Rettig who helped finish typing the manuscript on time.

The research performed by our group, which has been presented in this book, was supported by a contract from the US Department of Energy and a grant from the National Science Foundation.

SUBRATA MUKHERJEE

Contents

CHAPTER 1

Introduction

The boundary element method (BEM—also called the boundary integral equation method) is a powerful general-purpose procedure for the solution of boundary value problems in many branches of science and engineering. This method is based on an integral equation formulation of a given problem and has been used to solve problems in finite regions of arbitrary geometry as well as in infinite regions. The method was initially applied to linear problems but several nonlinear problems have been attacked by the method in recent years.

The boundary element method has several potential advantages over the widely used finite element method (FEM) for the solution of boundary value problems. One advantage is that the number of unknowns in resultant algebraic systems depends only on the boundary (or surface) discretization in BEM rather than on the discretization of the entire domain of the body as in FEM. The resultant matrices from the BEM are fully populated but tend to be numerically well conditioned. This arises from the fact that the singular kernels in the integral equations weigh the unknown quantities near a singular point more heavily than those far from it, thus causing the dominant components in a coefficient matrix to lie on or near its diagonal.

Another advantage of the BEM is that physical quantities obtained by differentiation of the primary variables, for example stresses or curvatures obtained from displacements, are determined pointwise inside and on the body, so that discontinuities in these variables across element boundaries cannot arise. This can be very important in problems of inelastic deformation as will be discussed later in the book. A third important advantage is that problems in infinite regions can be solved as easily as those in bounded regions. Finally, in problems where internal discretization is necessary in order to evaluate integrals with known integrands over the domain of a body, the topology of the internal cells is much simpler than for the FEM. In fact, the internal cells need not even cover the whole domain in some examples, as is the case with inelastic fracture mechanics problems discussed in the last two chapters of the book.

An important limitation of the straightforward BEM formulation is that the singular kernels of the integral equations must be known in an appropriate infinite body having the same material properties as the finite body being considered in a given problem. While these fundamental solutions are easily available in the applied mathematics literature for certain differential operators in homogeneous media, they might be difficult to obtain for a body composed of a general nonhomogeneous material. It is possible, however, to solve certain problems in nonhomogeneous media, by iterative methods using the kernel(s) for the homogeneous medium.

The heart of the boundary element method, as mentioned earlier, is the integral equation formulation of a given boundary value problem. The mathematical basis of this approach, of course, is classical and numerous applications of Green's functions have been reported in the literature. The earliest computer applications of the method date back about two decades and include the areas of potential theory,[1,2] fluid mechanics[3,4] and wave scattering.[5] These papers might be said to have heralded the modern era of the method.

Applications of the method have gained considerable momentum in recent years, in step, it appears, with the rapid improvements in computers. In the general area of solid mechanics, the method has been applied to a large class of linear problems. This research spans most of the important subject areas within solid mechanics, namely, linear elasticity,[6,7] linear viscoelasticity,[8] thermoelasticity,[9] linear elastic fracture mechanics,[10,11] elastic torsion,[12] bending of elastic plates,[13] shell theory[14] and wave propagation in elastic media.[15-17] This reference list is by no means complete, but is intended to be a collection of titles of some of the early papers in each subject area.

Applications of the BEM to nonlinear problems in solid mechanics is of more recent vintage, with the first formulation for time-independent plasticity being that due to Swedlow and Cruse.[18] This was followed by a numerical implementation by Riccardella.[19] Another formulation for plasticity using an equivalent body force based on initial stress is due to Banerjee and Mustoe.[20] The author of this book, together with his coauthors, have been active in the applications of the BEM to nonlinear problems of time-dependent inelastic deformation. A BEM formulation for viscoplastic problems[21] was followed by a numerical implementation for planar problems.[22] Other applications include inelastic torsion of prismatic shafts,[23] inelastic bending of plates,[24] and, very recently, inelastic fracture mechanics.[25-26] Currently, there is a great deal of activity in computer applications of the BEM to a wide class of problems in engineering science. The reader is referred to three recent books[27-29] which summarize the current state of the art of BEM applications in excellent fashion.

The purpose of this book is the presentation of applications of the boundary element method to nonlinear problems of time-dependent inelastic deformation and inelastic fracture mechanics. A discussion of constitutive models for inelastic deformation (creep and combined creep–plasticity or viscoplasticity) is given in Chapter 2. General formulations for three-dimensional problems (Chapter 3) are followed by a discussion of solution strategy and time integration in Chapter 4. Time integration with automatic time-step selection plays a crucial role in the successful completion of solutions of these time-dependent problems. Specific numerical applications to planar problems (Chapter 5), axisymmetric problems (Chapter 6), inelastic torsion (Chapter 7) and bending of thin plates (Chapter 8) follow. In each of these cases, comparisons with the results of finite element method solutions, and, whenever possible, comparisons with direct solutions from finite difference type techniques, have been carried out. The last two chapters of the book are concerned with applications of the method to problems of inelastic fracture mechanics. The time-histories of stresses near the tip of a stationary crack in a plate undergoing anti-plane shear or planar creep deformation are calculated, and the results are compared with recent asymptotic analytical solutions. In essence, this book is an attempt to present a comprehensive and up-to-date account of BEM applications in time-dependent inelastic problems in solid mechanics and to demonstrate the power of the method in solving these complicated nonlinear problems.

REFERENCES

1. Jaswon, M. A. Integral equation methods in potential theory, I. *Proceedings of the Royal Society*, London, Series A, **275**, 23–32 (1963).
2. Symm, G. T. Integral equation methods in potential theory, II. *Proceedings of the Royal Society*, London, Series A, **275**, 33–46 (1963).

3. HESS, J. *Calculation of Potential Flow about Arbitrary Three-Dimensional Bodies*. Douglas Aircraft Company Report ES 40622 (1962).
4. HESS, J. Calculation of potential flow about bodies of revolution having axes perpendicular to the free stream direction. *Journal of the Aeronautical Sciences*, **29**, 726–42 (1962).
5. FRIEDMAN, M. E. and SHAW, R. P. Diffraction of a plane shock wave by an arbitrary rigid cylindrical obstacle. *American Society of Mechanical Engineers, Journal of Applied Mechanics*, **29**, 40–46 (1962).
6. RIZZO, F. J. An integral equation approach to boundary value problems of classical elastostatics. *Quarterly of Applied Mathematics*, **25**, 83–95 (1967).
7. CRUSE, T. A. Numerical solutions in three dimensional elastostatics, *International Journal of Solids and Structures*, **5**, 1259–1274 (1969).
8. RIZZO, F. J. and SHIPPY, D. J. An application of the correspondence principle of linear viscoelasticity theory. *Society of Industrial and Applied Mathematics, Journal of Applied Mathematics*, **21**, 321–330 (1971).
9. RIZZO, F. J. and SHIPPY, D. J. An advanced boundary integral equation method for three-dimensional thermoelasticity. *International Journal for Numerical Methods in Engineering*, **11**, 1753–1768 (1977).
10. SNYDER, M. D. and CRUSE, T. A. Boundary integral equation analysis of cracked anisotropic plates. *International Journal of Fracture*, **11**, 315–328 (1975).
11. CRUSE, T. A. *Boundary-Integral Equation Method for Three-dimensional Elastic Fracture Mechanics Analysis*, Air Force Office of Scientific Research—TR–75–0813, Accession No. ADA 011660 (1975).
12. JASWON, M. A. and PONTER, A. R. An integral equation solution of the torsion problem, *Proceedings of the Royal Society*, London, Series A, **273**, 237–246 (1963).
13. JASWON, M. A. and MAITI, M. An integral equation formulation of plate bending problems. *Journal of Engineering Mathematics*, **2**, 83–93 (1968).
14. NEWTON, D. A. and TOTTENHAM, H. Boundary value problems in thin shallow shells of arbitrary plan form. *Journal of Engineering Mathematics*, **2**, 211–224 (1968).
15. CRUSE, T. A. and RIZZO, F. J. A direct formulation and numerical solution of the general transient elastodynamic problem I, *Journal of Mathematical Analysis and Applications*, **22**, 244–259 (1968).
16. CRUSE, T. A. A direct formulation and numerical solution of the general transient elastodynamic problem II, *Journal of Mathematical Analysis and Applications*, **22**, 341–355 (1968).
17. SHAW, R. P. Retarded potential approach to the scattering of elastic waves by rigid obstacles of arbitrary shape. *Journal of the Acoustical Society of America*, **44**, 745–748 (1968).
18. SWEDLOW, J. L. and CRUSE, T. A. Formulation of boundary integral equations for three-dimensional elasto-plastic flow. *International Journal of Solids and Structures*, **7**, 1673–1681 (1971).
19. RICCARDELLA, P. *An Implementation of the Boundary Integral Technique for Plane Problems of Elasticity and Elasto-Plasticity*. PhD Thesis, Carnegie Mellon University, Pittsburg, PA (1973).
20. BANERJEE, P. K. and MUSTOE, G. C. W. The boundary element method for

two-dimensional problems of elasto-plasticity. *Recent Advances in Boundary Element Methods*, C. A. Brebbia (ed.), Pentech Press, Plymouth, Devon, UK, 283–300 (1978).

21. KUMAR, V. and MUKHERJEE, S. A boundary-integral equation formulation for time-dependent inelastic deformation in metals. *International Journal of Mechanical Sciences*, **19**, 713–724 (1977).
22. MUKHERJEE, S. and KUMAR, V. Numerical analysis of time-dependent inelastic deformation in metallic media using the boundary-integral equation method. *American Society of Mechanical Engineers, Journal of Applied Mechanics*, **45**, 785–790 (1978).
23. MUKHERJEE, S. and MORJARIA, M. Comparison of boundary element and finite element methods in the inelastic torsion of prismatic shafts. *International Journal for Numerical Methods in Engineering*, **17**, 1576–1588 (1981).
24. MORJARIA, M. and MUKHERJEE, S. Inelastic analysis of transverse deflection of plates by the boundary element method. *American Society of Mechanical Engineers Journal of Applied Mechanics*, **47**, 291–296 (1980).
25. MUKHERJEE, S. and MORJARIA, M. A boundary element formulation for planar time-dependent inelastic deformation of plates with cutouts. *International Journal of Solids and Structures*, **17**, 115–126 (1981).
26. MORJARIA, M. and MUKHERJEE, S. Numerical analysis of planar, time-dependent inelastic deformation of plates with cracks by the boundary element method. *International Journal of Solids and Structures*, **17**, 127–143 (1981).
27. BANERJEE, P. K. and BUTTERFIELD, R. (eds.), *Developments in Boundary Element Methods*—1, Applied Science Publishers Ltd, Barking, Essex, UK (1979).
28. BANERJEE, P. K. and SHAW, R. P. (eds.), *Developments in Boundary Element Methods—2*, Applied Science Publishers Ltd, Barking, Essex, UK (1982).
29. BANERJEE, P. K. and BUTTERFIELD, R. *Boundary Element Methods in Engineering Science*, McGraw Hill, UK (1981).

CHAPTER 2

Constitutive Models

Constitutive models for the description of material behavior in the inelastic regime are discussed in this chapter. The materials considered are assumed to be metallic and the displacements and strains are assumed to remain small enough so that no distinction needs to be made between initial and current configurations.

2.1 CONSTITUTIVE MODELS FOR CREEP

Conventional design and analysis of metallic structures undergoing time-dependent inelastic deformation is generally carried out by linearly decomposing the total strain ε_{ij} into elastic ($\varepsilon_{ij}^{(e)}$), creep ($\varepsilon_{ij}^{(c)}$), plastic ($\varepsilon_{ij}^{(pl)}$) and thermal ($\varepsilon_{ij}^{(T)}$) components and then using separate constitutive descriptions for each of these components. Thus

$$\varepsilon_{ij}=\varepsilon_{ij}^{(e)}+\varepsilon_{ij}^{(c)}+\varepsilon_{ij}^{(pl)}+\varepsilon_{ij}^{(T)} \tag{2.1}$$

Hooke's law is used to relate the elastic strains and stresses σ_{ij}

$$\varepsilon_{ij}^{(e)}=\frac{1}{E}\{(1+\nu)\sigma_{ij}-\nu\sigma_{kk}\delta_{ij}\} \tag{2.2}$$

where E is the Young's modulus, ν is the Poisson's ratio, δ_{ij} is the Kronecker delta and the summation convention is used over the repeated index k. The thermal strain is generally written as

$$\varepsilon_{ij}^{(T)}=\alpha T\delta_{ij} \tag{2.3}$$

in terms of the temperature (above some base temperature) T and the coefficient of linear thermal expansion α. The plastic strain is generally described in terms of a yield criterion, flow rule and hardening law (see, for example, reference 1). The creep strain is generally assumed to depend on the stress, temperature and time (or strain) and is generally written in rate form in order to include, in some measure, the history dependence of the creep process. Based on an expression of the type

$$\varepsilon^{(c)} = A\sigma^n t^\mu \tag{2.4}$$

which is assumed to be valid for *uniaxial* creep under *constant* stress and temperature (with A, n (usually > 1) and μ (usually < 1) parameters which depend on temperature and time 't'), several models for the rate of uniaxial creep strain have been postulated.[2] Some of the commonly used laws are

$$\text{Power law creep: } \dot{\varepsilon}^{(c)} = \dot{\varepsilon}_c \left(\frac{\sigma}{\sigma_c}\right)^n \tag{2.5}$$

$$\text{Time hardening creep: } \dot{\varepsilon}^{(c)} = \dot{\varepsilon}_c \left(\frac{\sigma}{\sigma_c}\right)^n \mu t^{\mu-1} \tag{2.6}$$

$$\text{Strain hardening creep: } \dot{\varepsilon}^{(c)} = \dot{\varepsilon}_c^{1/\mu} \left(\frac{\sigma}{\sigma_c}\right)^{n/\mu} \mu(\varepsilon^{(c)})^{(\mu-1)/\mu} \tag{2.7}$$

In the above $\dot{\varepsilon}_c$ and σ_c are reference strain rate and stress, respectively, and can be functions of temperature. Each of these laws is assumed to be valid for variable stress situations.

A multiaxial generalization is obtained in the form of a flow rule

$$\dot{\varepsilon}_{ij}^{(c)} = \frac{3}{2}\frac{\dot{\varepsilon}^{(c)}}{\sigma} s_{ij} \tag{2.8}$$

where $s_{ij} = \sigma_{ij} - \frac{1}{3}\sigma_{kk}\delta_{ij}$ is the deviatoric stress tensor and $\dot{\varepsilon}^{(c)}$ and σ are strain rate and stress invariants, respectively, defined as

$$\sigma = \sqrt{\frac{3}{2} s_{ij} s_{ij}}, \quad \dot{\varepsilon}^{(c)} = \sqrt{\frac{2}{3} \dot{\varepsilon}_{ij}^{(c)} \dot{\varepsilon}_{ij}^{(c)}} \tag{2.9}$$

A uniaxial equation of the type (2.5) is now assumed to retain its form with the uniaxial quantities replaced by the invariants of eqn. (2.9). While this multiaxial formulation is not unique, it has several desirable features. It is consistent with an appropriate uniaxial equation and reproduces certain experimentally observed features like creep deformation being volume preserving and the hydrostatic component of stress having no effect on creep.

A model like eqn. (2.7), or a modified version thereof, is commonly used in the design of structures in creep. However, researchers like Krempl[3,4] and Onat and Fardshisheh[5] have critically examined these theories and have concluded that they are incapable of representing all the salient features of high-temperature inelastic deformation of metals. For example, the strain hardening and time hardening theories of creep do not adequately take into account the effect of prior deformation history on subsequent creep behavior, and both are incapable of representing a softening of the material which accompanies creep recovery. Further, these laws generally do not predict material behavior in the presence of stress reversals in a satisfactory manner. This has led to the search for a new generation of models to describe inelastic deformation in an accurate manner for a wide class of stress and temperature histories. Some of these models are discussed in the next section.

2.2 NEW GENERATION OF CONSTITUTIVE MODELS FOR INELASTIC DEFORMATION

A new generation of constitutive models has been proposed during the past decade with the goal of overcoming the shortcomings of the models described previously and of predicting material response in the inelastic regime in a more accurate manner than is possible with the traditional models. Some of these papers are listed as references.[6–17] Many (but not all) of these theories combine plastic and creep strains into a single unified nonelastic strain ($\varepsilon_{ij}^{(n)}$) and often include certain internal or state variables to define the current deformation state of an element in a structure. A subset of these models has the mathematical structure given below, and it is this subset that is of concern in this book. This class of models can be summarized by the following equations

$$\dot{\varepsilon}_{ij}=\dot{\varepsilon}_{ij}^{(e)}+\dot{\varepsilon}_{ij}^{(n)}+\dot{\varepsilon}_{ij}^{(T)} \tag{2.10}$$

$$\dot{\varepsilon}_{ij}^{(n)}=f_{ij}(\sigma_{ij},q_{ij}^{(k)},T) \tag{2.11}$$

$$\dot{q}_{ij}^{(k)}=g_{ij}(\sigma_{ij},q_{ij}^{(k)},T) \tag{2.12}$$

$$\dot{\varepsilon}_{kk}^{(n)}=0 \tag{2.13}$$

where the new variables, $q_{ij}^{(k)}$, are state variables. The number of state variables varies in the different models (usually one or two are used in a given model) and they can be scalars or tensors. These are often physically motivated by micromechanical considerations like dislocation density and resistance to dislocation motion. The models, however, are phenomenological in nature, and the state variables, in mechanics terminology, are analogous to quantities like current yield stress, kinematic hardening parameters or accumulated plastic work. A small number of these state variables are assumed completely to characterize the present deformation state of a material, and the history dependence of the rate of the nonelastic strain up to the current time is assumed to be completely represented by their current values. It is important to note that the rates of the nonelastic strain components and the state variables at any time depend only on the current values of the stress, state variables and temperature, in an explicit manner.

The mathematical structure of conventional models for creep (eqns. (2.5)–(2.9)) fit into this general format provided $\dot{\varepsilon}_{ij}^{(n)}$ is interpreted as the rate of creep strain. The conventional separate treatment of plastic strain (eqn. (2.1)), however, does not, since, according to standard plasticity theory in the presence of work hardening, a plastic strain increment is a function of the current stress as well as the stress increment.[1] A constitutive model with the mathematical structure described by eqns. (2.10)–(2.13), in addition to being an improvement over a conventional theory for the purpose of material modelling, also has the advantage of being easier to use in a BEM or FEM algorithm for the solution of boundary value problems. This fact will be elaborated upon in subsequent chapters of this book.

2.3 THE CONSTITUTIVE MODEL DUE TO HART

2.3.1 The Model

The constitutive model due to Hart[10,11] has been extensively tested for uniaxial time-varying loading on various metals and alloys. Many of these results, with comparisons of prediction and experiment, are available in a recent Electric Power Research Institute report.[18] Some comparisons of theory and experiment on bending of beams have also been completed[19] and multiaxial experiments at room and elevated temperatures are currently in progress at Cornell. The correlation between theory and experiment has generally been found to be very good. This constitutive model is used in many of the numerical calculations presented later in this chapter. A brief review of the model is presented next.

The model in uniaxial tension is shown in Fig. 2.1. Material response in dilatation is assumed to be elastic. According to this model, the nonelastic strain is decomposed into two (time-dependent) components

$$\dot{\varepsilon}_{ij}^{(n)}=\dot{\varepsilon}_{ij}^{(a)}+\dot{\varepsilon}_{ij}^{(p)} \tag{2.14}$$

where $\varepsilon_{ij}^{(a)}$ is the anelastic strain, a stored strain which reflects the magnitude and direction of prior deformation history and $\varepsilon_{ij}^{(p)}$ is the completely irrecoverable and path dependent permanent strain. The two state variables in the model are the anelastic strain and a scalar σ^*, called hardness, which is similar to an isotropic strain hardening parameter or current yield stress.

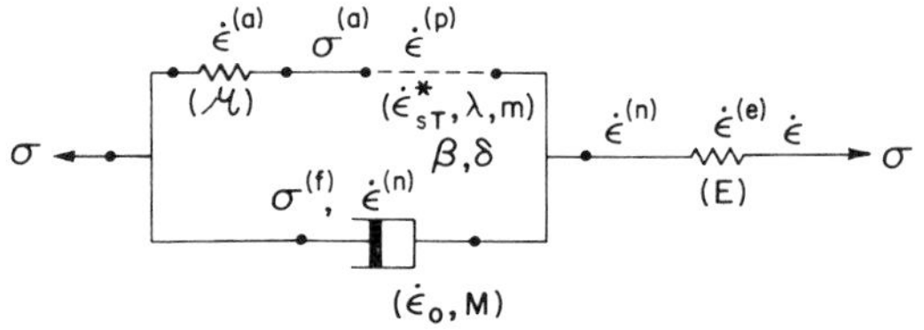

FIG. 2.1. Hart's model in tension.

The deviatoric component of the stress, s_{ij}, is decomposed into two auxiliary tensors

$$s_{ij}=s_{ij}^{(a)}+s_{ij}^{(f)} \tag{2.15}$$

The flow rules for the strains or strain rates are of the form

$$\varepsilon_{ij}^{(a)}=\frac{3}{2}\frac{\varepsilon^{(a)}}{\sigma^{(a)}}s_{ij}^{(a)} \tag{2.16}$$

$$\dot{\varepsilon}_{ij}^{(p)}=\frac{3}{2}\frac{\dot{\varepsilon}^{(p)}}{\sigma^{(a)}}s_{ij}^{(a)} \tag{2.17}$$

$$\dot{\varepsilon}_{ij}^{(n)}=\frac{3}{2}\frac{\dot{\varepsilon}^{(n)}}{\sigma^{(f)}}s_{ij}^{(f)} \tag{2.18}$$

where $\dot{\varepsilon}^{(n)}$, $\dot{\varepsilon}^{(p)}$, $\varepsilon^{(a)}$, σ, $\sigma^{(a)}$ and $\sigma^{(f)}$ are scalar invariants of the corresponding tensors, defined as

$$\begin{aligned}\dot{\varepsilon}^{(n)}&=\sqrt{\frac{2}{3}\dot{\varepsilon}_{ij}^{(n)}\dot{\varepsilon}_{ij}^{(n)}}, \quad \dot{\varepsilon}^{(p)}=\sqrt{\frac{2}{3}\dot{\varepsilon}_{ij}^{(p)}\dot{\varepsilon}_{ij}^{(p)}}, \quad \varepsilon^{(a)}=\sqrt{\frac{2}{3}\varepsilon_{ij}^{(a)}\varepsilon_{ij}^{(a)}}\\ \sigma&=\sqrt{\frac{3}{2}s_{ij}s_{ij}}, \quad \sigma^{(a)}=\sqrt{\frac{3}{2}s_{ij}^{(a)}s_{ij}^{(a)}}, \quad \sigma^{(f)}=\sqrt{\frac{3}{2}s_{ij}^{(f)}s_{ij}^{(f)}}\end{aligned} \tag{2.19}$$

These scalar invariants are related to each other through the uniaxial equations (see Fig. 2.1)

$$\sigma^{(a)}=\mathcal{M}\varepsilon^{(a)},\ \dot{\varepsilon}^{(n)}=\dot{\varepsilon}_0(\sigma^{(f)}/\sigma_0)^M \tag{2.20, 2.21}$$

$$\dot{\varepsilon}^{(p)}=\dot{\varepsilon}^*\{\ln(\sigma^*/\sigma^{(a)})\}^{-1/\lambda} \tag{2.22}$$

$$\dot{\varepsilon}^*=\dot{\varepsilon}^*_{sT}(\sigma^*/\sigma^*_s)^m\exp(Q/RT_B)\exp(-Q/RT) \tag{2.23}$$

$$\dot{\sigma}^*=\dot{\varepsilon}^{(p)}\sigma^*\Gamma(\sigma^*,\sigma^{(a)}) \tag{2.24}$$

$$\Gamma(\sigma^*,\sigma^{(a)})=(\beta/\sigma^*)^\delta(\sigma^{(a)}/\sigma^*)^{\beta/\sigma^*} \tag{2.25}$$

Equation (2.20) represents a linear anelastic element, (2.21) a nonlinear dashpot, (2.22) and (2.23) a 'plastic' element and, finally, (2.24) and (2.25) describe strain hardening. The flow parameters are $\mathcal{M}$, M, m, λ, $\dot{\varepsilon}_0$ (at a reference stress level σ_0) and $\dot{\varepsilon}^*_{sT}$ (at a reference hardness level σ^*_s and reference temperature T_B); β and δ are strain hardening parameters, R is the gas constant and Q the activation energy for atomic self diffusion.

2.3.2 The Viscoplastic Limit

The eqns. (2.15)–(2.25) can be directly used for numerical computation. However, in the region $\sigma^{(a)}\approx\sigma^*$ (analogous to plastic flow at the yield stress) eqn. (2.22) predicts large values of $\dot{\varepsilon}^{(p)}$ and consequently very small time steps are required in a numerical computation process. For the sake of computational efficiency, the condition $\dot{\sigma}^{(a)}=\dot{\sigma}^*$ is used for this region. Using this condition, it can be shown that the invariant for the permanent strain rate

$$\dot{\varepsilon}^{(p)}=\frac{s_{ij}^{(a)}\dot{\varepsilon}_{ij}^{(n)}/\sigma^{(a)}}{1+\sigma^*\Gamma/\mathcal{M}} \tag{2.26}$$

The region where $\sigma^{(a)}\approx\sigma^*$ and $\dot{\sigma}^a=\dot{\sigma}^*$ is called the viscoplastic limit and eqn. (2.26) replaces (2.22) in this region. This assumption can be justified experimentally.[18]

An algorithm for obtaining numerical results proceeds as follows. The initial value of σ^* is prescribed (usually the yield stress of the specimen) and usually $\sigma^{(a)}$ is initially taken to be zero. During computation, $\sigma^{(a)}$ grows and approaches σ^*. If $\sigma^{(a)}=\upsilon\sigma^*$, where υ is a number close to unity (usually 0·99—this is analogous to yielding and onset of plastic flow), eqn. (2.26) is used instead of (2.22). The permanent strain rate remains relatively large in the viscoplastic limit. If $\dot{\varepsilon}^{(p)}$ drops below a critical $\dot{\varepsilon}^{(p)}_c$, defined as

$$\dot{\varepsilon}_c^{(p)} = \dot{\varepsilon}^* \ln\left(\frac{1}{\upsilon}\right)^{-1/\lambda} \tag{2.27}$$

the computer program exits the viscoplastic limit and eqn. (2.22) is used again. The critical strain rate $\dot{\varepsilon}_c^{(p)}$ is the value of the permanent strain rate when $\sigma^{(a)} = \upsilon\sigma^*$. Further, if at any time there is a sharp discontinuity in loading history within the viscoplastic limit, eqn. (2.22) should replace (2.26).

2.3.3 Determination of Material Parameters

The material parameters for Hart's model are obtained from two kinds of basic experiments. The 'plastic' element parameters $\dot{\varepsilon}_{sT}^*$, λ and m and the nonlinear dashpot parameters M and $\dot{\varepsilon}_0$ are obtained from relaxation tests. The anelastic modulus $\mathcal{M}$ and the work hardening parameters β and δ are obtained from constant strain rate tension tests. Details of this procedure are available in reference 18. A brief outline of the procedure follows.

The uniaxial load relaxation test is carried out under constant strain, i.e. $\dot{\varepsilon} = 0$. It is assumed that the hardness σ^* remains constant during this test since the change in the structure of the specimen is minimal. Two situations are analysed separately.

The transient free case to get the flow parameters of the plastic element. In this case the applied stress is kept below the hardness of the specimen. It is assumed that

$$\sigma^* > \sigma = \sigma^{(a)}, \ \dot{\varepsilon}^{(p)} = \dot{\varepsilon}^{(n)}$$

i.e. the anelastic element is saturated so that $\dot{\varepsilon}^{(a)} \approx 0$. Now, $\dot{\varepsilon}^{(p)} = \dot{\varepsilon}^{(n)} = -\dot{\varepsilon}^{(e)} = -\dot{\sigma}/E$. Equation (2.21) is ignored and eqn. (2.22) reduces to

$$\ln(\sigma^*/\sigma) = (\dot{\varepsilon}^*/\dot{\varepsilon}^{(p)})^{\lambda}$$

Plots of $\ln\sigma$ as a function of $\ln\dot{\varepsilon}^{(p)}$ are obtained for different values of hardness. These curves follow the preceding equation, and, if translated rigidly along a line of slope $1/m$, coalesce into a single master curve. The relaxation experiments are repeated at different temperatures to obtain the temperature dependence of $\dot{\varepsilon}^*$ in eqn. (2.23). In this way, the flow parameters λ, m and $\dot{\varepsilon}_{sT}^*$ of the plastic element are obtained.

The high stress case to get the flow parameters of the nonlinear dash pot. In this case, the applied stress is kept above the hardness of the specimen.

It is assumed that

$$\sigma > \sigma^* = \sigma^{(a)}$$

i.e. the plastic element is deforming at the 'yield' stress σ^*.

Now, eqns. (2.15) and (2.21) give

$$\dot{\varepsilon}^{(n)} = \dot{\varepsilon}_0 \{(\sigma - \sigma^*)/\sigma_0\}^M$$

with, once again, $\dot{\varepsilon}^{(n)} = -\dot{\sigma}/E$.

Plots of $\ln\sigma$ versus $\ln\dot{\varepsilon}^{(n)}$ at different values of hardness are used once again, this time to obtain the flow parameters $\dot{\varepsilon}_0$ and M of the nonlinear dashpot.

The other basic experiment used to obtain material parameters is the uniaxial tension test at constant strain rate $\dot{\varepsilon}$. A plot of σ versus accumulated strain ε is obtained in this case. The initial portion of this curve, for low strain rate, shows a transition in slope from E to $\mathcal{M}E/(\mathcal{M}+E)$, corresponding to saturation of the anelastic spring but negligible deformation of the plastic element. This slope is used to obtain the anelastic modulus. Finally, the work hardening part of the curve, that for relatively large strains, is used to obtain the work hardening parameters β and δ. To do this, the experimentally obtained data from the tension test are first reduced to give curves for $\ln\sigma^*$ and $\ln\sigma^{(a)}$ as functions of the accumulated permanent strain $\varepsilon^{(p)}$. The parameters β and δ in eqn. (2.25) are obtained from these curves.

REFERENCES

1. MENDELSON, A. *Plasticity: Theory and Application*, Macmillan Company, New York (1968).
2. KRAUS, H. *Creep Analysis*, John Wiley and Sons, New York (1980).
3. KREMPL, E. The interaction of rate and history dependent effects and its significance for slow cyclic inelastic analysis at elevated temperatures. *Nuclear Engineering and Design*, **29**, 125–134 (1974).
4. KREMPL, E. Cyclic creep—an interpretive literature survey. *Welding Research Council Bulletin*, **195**, 63–123 (1974).
5. ONAT, E. T. and FARDSHISHEH, F. *Representation of Creep of Metals*. Oak Ridge National Laboratory Report 4783 (1972).
6. BODNER, S. R. and PARTOM, Y. Constitutive equations for elastic viscoplastic strain-hardening materials, American Society of Mechanical Engineers, *Journal of Applied Mechanics*, **42**, 385–389 (1975).
7. CERNOCKY, E. P. and KREMPL, E. A theory of thermoviscoplasticity based on infinitesimal total strain, *International Journal of Solids and Structures*, **16**, 723–741 (1980).

8. GITTUS, J. H. Development of constitutive relation for plastic deformation from a dislocation model, American Society of Mechanical Engineers *Journal of Engineering Materials and Technology*, **98**, 52–59 (1976).
9. GITTUS, J. H. Microstructure based modelling of constitutive behavior for engineering applications, *Constitutive Equations in Plasticity*, A. S. Argon (ed.), MIT Press, Cambridge, MA, 487–548 (1976).
10. HART, E. W., LI, C. Y., YAMADA, H. and WIRE, G. L. Phenomenological theory—a guide to constitutive relations and fundamental deformation properties, *Constitutive Equations in Plasticity*, A. S. Argon (ed.), MIT Press, Cambridge, MA, 149–197 (1976).
11. HART, E. W. Constitutive relations for the non-elastic deformation of metals, American Society of Mechanical Engineers *Journal of Engineering Materials and Technology*, **98**, 193–202 (1976).
12. KRIEG, R. D. A physically based internal variable model for rate-dependent plasticity, *Inelastic Behavior of Pressure Vessel and Piping Components*, T. Y. Chang and E. Krempl (eds.), American Society of Mechanical Engineers, New York, 15–28 (1978).
13. MILLER, A. An inelastic constitutive model for monotonic, cyclic and creep deformation: Part I—Equations Development and Analytical Procedures, American Society of Mechanical Engineers *Journal of Engineering Materials and Technology*, **98**, 97–105 (1976).
14. MILLER, A. An inelastic constitutive model for monotonic, cyclic and creep deformation, Part II—Application to type 304 stainless steel, American Society of Mechanical Engineers *Journal of Engineering Materials and Technology*, **98**, 106–113 (1976).
15. PONTER, A. R. S. and LECKIE, F. A. Constitutive relations for the time-dependent deformation of metals, American Society of Mechanical Engineers *Journal of Engineering Materials and Technology*, **98**, 47–51 (1976).
16. ROBINSON, D. N. *A Candidate Creep-Recovery Model for 2-1/4 Cr–1 Mo Steel and its Experimental Implementation.* Oak Ridge National Laboratory Report ORNL-TM-5110 (1975).
17. VALANIS, K. C. On the foundation of the endochronic theory of viscoplasticity, *Archives of Mechanics*, **27**, 857–868 (1975).
18. KUMAR, V., MUKHERJEE, S., HUANG, F. H. and LI, C-Y. *Deformation in Type 304 Austenitic Stainless Steel*, Electric Power Research Institute Report EPRI NP-1276 for Project 697-1, Palo Alto, CA (1979).
19. MAUTZ, J. and HART, E. W. *The Inelastic Bending of a Nickel Beam at Several Temperatures: Experiment and Prediction.* DOE report C00-2733-29, Department of Theoretical and Applied Mechanics, Cornell University, Ithaca, NY (1980).

CHAPTER 3

Three-dimensional Formulations

Three-dimensional boundary element formulations for elasticity, time-independent plasticity and time-dependent viscoplastic deformation are presented in this chapter. The range of subscripts in all the equations in this chapter is 1, 2, 3.

3.1 ELASTICITY

3.1.1 Governing Differential Equations

The governing differential equations[1] for linear elasticity, for a homogeneous, isotropic, three-dimensional body, are the equilibrium equations

$$\sigma_{ij,j} = -F_i \tag{3.1}$$

the kinematic equations (for small strains and displacements)

$$\varepsilon_{ij}^{(e)} = \frac{1}{2}(u_{i,j} + u_{j,i}) \tag{3.2}$$

and the constitutive equations (generalized Hooke's law)

$$\sigma_{ij} = \lambda \varepsilon_{kk}^{(e)} \delta_{ij} + 2G\varepsilon_{ij}^{(e)} \tag{3.3}$$

In the above equations, u_i are the components of the displacement vector, F_i are the components of the body force vector per unit volume, and λ and G are Lame constants. A comma denotes differentiation with respect to the appropriate spatial variable in the usual way.

The above equations can be combined to yield the Navier equations for the displacement

$$u_{i,jj} + \frac{1}{1-2\nu} u_{k,ki} = -\frac{F_i}{G} \tag{3.4}$$

The traction at a point on the surface ∂B of a body B is

$$\tau_i = \sigma_{ij} n_j = G\left\{ (u_{i,j} + u_{j,i}) n_j + \frac{2\nu}{1-2\nu} u_{k,k} n_i \right\} \tag{3.5}$$

where n_i are the components of the unit outward normal to ∂B at that point.

The usual boundary conditions require specification of tractions over a part of the boundary and displacements over the rest of the boundary ∂B, or, for mixed-mixed problems, appropriate components of these variables over ∂B.

3.1.2 Boundary Element Formulation

A boundary element formulation for elasticity can be based on the reciprocal theorem of Betti[2]

$$\int_{\partial B} u_i \tau_i^* \, \mathrm{d}s + \int_B u_i F_i^* \, \mathrm{d}v = \int_{\partial B} u_i^* \tau_i \, \mathrm{d}s + \int_B u_i^* F_i \, \mathrm{d}v \tag{3.6}$$

where (F_i^*, u_i^*, τ_i^*) are the body forces, displacements and tractions due to a point force in an infinite three-dimensional elastic solid and (F_i, u_i, τ_i) are the corresponding quantities in the problem of interest. Thus,

$$F_i^* = \Delta(p,q)\delta_{ij} e_j \tag{3.7}$$

$$u_i^* = U_{ij} e_j \tag{3.8}$$

$$\tau_i^* = T_{ij} e_j \tag{3.9}$$

where p is a source point, q is a field point and $\Delta(p,q)$ is the Dirac delta function. The two-point function U_{ij} is the displacement at q in the i direction due to a unit point load at p in the j direction. The function T_{ij} has a similar physical meaning in terms of tractions. Unit orthogonal base vectors are denoted by e_j. These functions are known from Kelvin's singular solution due to a point load in an infinite elastic solid

$$U_{ij} = \frac{1}{16\pi(1-\nu)Gr}\{(3-4\nu)\delta_{ij} + r_{,i} r_{,j}\} \tag{3.10}$$

$$T_{ij} = -\frac{1}{8\pi(1-\nu)r^2}\left[\{(1-2\nu)\delta_{ij} + 3 r_{,i} r_{,j}\}\frac{\partial r}{\partial n} + (1-2\nu)(r_{,i} n_j - r_{,j} n_i) \right] \tag{3.11}$$

where $r(p,q)$ is the distance from a source point p to a field point q and n_i are the components of the unit outward normal to ∂B at a point Q on it. The convention used here is that lowercase letters p and q denote points inside the body B and capital letters denote points on its boundary ∂B. A comma denotes a derivative with respect to a field point, i.e.

$$r_{,i} = \frac{\partial r}{\partial x_{0i}} = \frac{x_{0i} - x_i}{r},$$

where x and x_0 are the source and field points, respectively.

Substituting eqns. (3.7)–(3.9) into (3.6) leads to the equation

$$u_j(p) = \int_{\partial B} \{U_{ij}(p,Q)\tau_i(Q) - T_{ij}(p,Q)u_i(Q)\} \mathrm{d}s_Q + \int_B U_{ij}(p,q)F_i(q)\,\mathrm{d}v_q \tag{3.12}$$

The surface integral in eqn. (3.12) cannot be immediately evaluated since both the traction and displacement fields are not known *a priori* over the entire surface ∂B. Thus, a limit of eqn. (3.12) must be taken as an internal point p approaches a boundary point P on ∂B. This limit must be taken carefully by excluding a sector of a circle around P and then shrinking this circle to a point.[3] This limiting process, for a point P on the boundary where it is not necessarily locally smooth, leads to the integral equation

$$c_{ij}(P)u_i(P) = \int_{\partial B} \{U_{ij}(P,Q)\tau_i(Q) - T_{ij}(P,Q)u_i(Q)\} \mathrm{d}s_Q + \int_B U_{ij}(P,q)F_i(q)\,\mathrm{d}v_q \tag{3.13}$$

The surface integrals in eqn. (3.13) must be interpreted in the sense of Cauchy principal values. The coefficients of c_{ij} depend on the local geometry of ∂B at P. If the boundary is locally smooth at P, $c_{ij} = (1/2)\delta_{ij}$. Otherwise, c_{ij} can be evaluated in closed form for two-dimensional problems,[4] but direct evaluation of c_{ij} in three dimensions is difficult. A convenient indirect approach is the imposition of rigid body displacements in a body with no body forces. This must give a stress-free state and leads to the auxiliary equation[5,6]

$$c_{ij}(P) + \int_{\partial B} T_{ij}(P,Q)\,\mathrm{d}s_Q = 0 \tag{3.14}$$

Equation (3.13) is a set of integral equations for the unknown components of the boundary tractions and displacements in terms of the prescribed ones. (The body forces are assumed to be prescribed.) It can be solved to yield both the traction and displacement fields over the entire boundary ∂B and these can then be used to obtain the displacement field throughout the body from eqn. (3.12). The strain and stress fields must be determined next.

The strain field throughout the body must be obtained by differentiating the displacement field using finite difference or finite element methods. Another way is to analytically differentiate eqn. (3.12). This requires differentiation of the kernels under the integral signs. The second approach is convenient[3] and leads to the equation

$$u_{j,L}(p)=\int_{\partial B}\{U_{ij,L}(p,Q)\tau_i(Q)-T_{ij,L}(p,Q)u_i(Q)\}\mathrm{d}s_Q$$

$$+\int_B U_{ij,L}(p,q)F_i(q)\,\mathrm{d}v_q \tag{3.15}$$

where a capital letter following a comma denotes differentiation with respect to a source point. Using the identity

$$r_{,L}=-r_{,l}$$

the differentiated kernels can be written in terms of field point derivatives as

$$U_{ij,L}=-U_{ij,l}=\frac{1}{16\pi G(1-\nu)r^2}\{-r_{,i}\delta_{jl}-r_{,j}\delta_{li}+(3-4\nu)r_{,l}\delta_{ij}+3r_{,i}r_{,j}r_{,l}\} \tag{3.16}$$

$$T_{ij,L}=\frac{1}{8\pi(1-\nu)r^3}\left[3(r_{,i}\delta_{jl}+r_{,j}\delta_{li}-5r_{,l}r_{,i}r_{,j})\frac{\partial r}{\partial n}+3r_{,i}r_{,j}n_l\right.$$

$$+(1-2\nu)\left\{\delta_{ij}n_l-\delta_{jl}n_i+\delta_{li}n_j\right.$$

$$\left.\left.+3\left(n_ir_{,j}r_{,l}-n_jr_{,l}r_{,i}-r_{,l}\delta_{ij}\frac{\partial r}{\partial n}\right)\right\}\right] \tag{3.17}$$

The kernel in the volume integral in eqn. (3.15) becomes singular when q coincides with p. This integral must be evaluated carefully. Equation

(3.15) is not valid for a boundary point ($p \to P$) since the differentiated kernels become singular and residues must be included. These residues are difficult to evaluate. In fact, even though eqn. (3.15) is correct for all internal points p, numerical errors in the displacement gradients can arise when p gets very close to the boundary (in the so-called 'boundary layer') due to the r^{-3} dependence of $T_{ij,L}$.[3] Thus, stresses on the boundary are best evaluated by an alternative method which is described next. For internal points sufficiently away from the boundary, the displacement gradients can be obtained from eqn. (3.15) and it is then a simple matter to determine the strains and stresses from the kinematic equations (3.2) and Hooke's law (3.3).

3.1.3 Boundary Stresses

The goal here is to determine the stress components at a point P on the boundary ∂B purely from boundary data. This can be achieved once eqn. (3.13) is solved to yield the displacement and traction fields over the entire boundary ∂B. This calculation involves the numerical determination of the displacement derivatives in a direction tangential to the boundary and the use of some of the laws of elasticity.[7] Any convenient direction tangential to the two-dimensional surface ∂B at P can be chosen for this purpose.

The calculation involves four vectors and two tensors at P. The vectors are the traction $\boldsymbol{\tau}$, the tangential derivative of the displacement $\partial \mathbf{u}/\partial s$, and the unit normal and tangent vectors $\mathbf{n}$ and $\mathbf{t}$ at P. The tensors are the stress $\boldsymbol{\sigma}$ and the displacement gradient $\partial \mathbf{u}/\partial \mathbf{x}$. Once eqn. (3.13) is solved and $\partial \mathbf{u}/\partial s$ evaluated numerically, these four vectors (twelve scalar components) are known at P. The twelve unknown tensor components σ_{ij} and $u_{i,j}$ can now be determined from the following set of three equations

$$\sigma_{ij} = \lambda u_{k,k} \delta_{ij} + G(u_{i,j} + u_{j,i}) \tag{3.18}$$

$$\tau_i = \sigma_{ij} n_j \tag{3.19}$$

$$\frac{\partial u_i}{\partial s} = u_{i,j} t_j \tag{3.20}$$

The result is the determination of all the stress and strain components at a point on the boundary ∂B. The stresses and strains in the 'boundary layer' can now be obtained by interpolation between the boundary and internal quantities.

3.2 TIME-INDEPENDENT PLASTICITY

3.2.1 Governing Differential Equations

It is convenient to write the plasticity equations in incremental or rate form. The kinematic equations, in the presence of 'small' plastic strains $\varepsilon_{ij}^{(pl)}$, are

$$\dot{\varepsilon}_{ij}=\dot{\varepsilon}_{ij}^{(e)}+\dot{\varepsilon}_{ij}^{(pl)}=\frac{1}{2}(\dot{u}_{i,j}+\dot{u}_{j,i}) \tag{3.21}$$

where a superposed dot denotes an increment (or a derivative with respect to a suitable loading parameter) of the appropriate quantity. The equilibrium equations (3.1) and Hooke's law (3.3) remain the same as before. Writing these in rate form and combining them with eqn. (3.21) gives the Navier equations for the displacement rate

$$\dot{u}_{i,jj}+\frac{1}{1-2v}\dot{u}_{k,ki}=-\frac{\dot{F}_i}{G}+2\dot{\varepsilon}_{ij,j}^{(pl)} \tag{3.22}$$

where the equation $\dot{\varepsilon}_{kk}^{(pl)}=0$ has been used. If desired, this term can be retained in the formulation without difficulty.

The traction rate on the surface of the body ∂B is given by

$$\dot{\tau}_i=\dot{\sigma}_{ij}n_j=G\left\{(\dot{u}_{i,j}+\dot{u}_{j,i})n_j+\frac{2v}{1-2v}\dot{u}_{k,k}n_i-2\dot{\varepsilon}_{ij}^{(pl)}n_j\right\} \tag{3.23}$$

The boundary conditions must prescribe traction or displacement histories on ∂B in the usual way.

A pseudo body force F_i' and a pseudo boundary traction τ_i' can be defined as

$$\dot{F}_i'=\dot{F}_i-2G\dot{\varepsilon}_{ij,j}^{(pl)} \tag{3.24}$$

$$\dot{\tau}_i'=\dot{\tau}_i+2G\dot{\varepsilon}_{ij}^{(pl)}n_j \tag{3.25}$$

and these reduce eqns. (3.22) and (3.23) to the same form as those for linear elasticity with pseudo body and boundary forces. Thus

$$\dot{u}_{i,jj}+\frac{1}{1-2v}\dot{u}_{k,ki}=-\frac{\dot{F}_i'}{G} \tag{3.26}$$

and

$$\frac{\dot{\tau}_i'}{G}=(\dot{u}_{i,j}+\dot{u}_{j,i})n_j+\frac{2v}{1-2v}\dot{u}_{k,k}n_i \tag{3.27}$$

3.2.2 Boundary Element Formulation for Displacement Rates

A boundary element formulation for plasticity can be based on the reciprocal theorem of Betti in terms of modified surface traction and displacement rates[8,9]

$$\int_{\partial B} \dot{u}_i \tau_i^* \mathrm{d}s + \int_B \dot{u}_i F_i^* \mathrm{d}v = \int_{\partial B} u_i^* \dot{\tau}_i' \mathrm{d}s + \int_B u_i^* \dot{F}_i' \mathrm{d}v \tag{3.28}$$

where, as before, (F_i^*, u_i^* and τ_i^*) are the body forces, displacements and tractions due to a point force in an infinite *elastic* solid and ($\dot{F}_i'$, $\dot{u}_i$, $\dot{\tau}_i'$) are the corresponding quantities in an elastic–plastic solid as defined above.

Substituting eqns. (3.7)–(3.9) into eqn. (3.28) leads to the equation

$$\dot{u}_j(p) = \int_{\partial B} \{U_{ij}(p,Q)\dot{\tau}_i'(Q) - T_{ij}(p,Q)\dot{u}_i(Q)\}\mathrm{d}s_Q + \int_B U_{ij}(p,q)\dot{F}_i'(q)\mathrm{d}v_q \tag{3.29}$$

A more convenient form of this equation can be obtained by substituting for $\dot{F}_i'$ and $\dot{\tau}_i'$ in terms of physical quantities from eqns. (3.24) and (3.25) and using the divergence theorem to transform some volume integrals into surface integrals. In the absence of physical body forces ($\dot{F}_i = 0$), this leads to

$$\dot{u}_j(p) = \int_{\partial B} \{U_{ij}(p,Q)\dot{\tau}_i(Q) - T_{ij}(p,Q)\dot{u}_i(Q)\}\mathrm{d}s_Q + \int_B 2G U_{ij,k}(p,q)\dot{\varepsilon}_{ik}^{(\mathrm{pl})}(q)\mathrm{d}v_q \tag{3.30}$$

This equation only involves physical quantities and does not have a spatial derivative of the plastic strain increment in the volume integral. The comma in the last term of eqn. (3.30) denotes, as usual, a derivative with respect to a field point q. The explicit form of the kernel in the volume integral, $U_{ij,k}$, is given by eqn. (3.16) (with the free index l replaced by k).

As before, an integral equation for the unknown components of the traction and displacement rates in terms of the prescribed ones can be obtained by taking the limit as p approaches a boundary point P. This leads to the equation

$$c_{ij}(P)\dot{u}_i(P)=\int_{\partial B}\{U_{ij}(P,Q)\dot{\tau}_i(Q)-T_{ij}(P,Q)\dot{u}_i(Q)\}\mathrm{d}s_Q + \int_B 2GU_{ij,k}(P,q)\dot{\varepsilon}_{ik}^{(\mathrm{pl})}(q)\mathrm{d}v_q \qquad (3.31)$$

The integral eqn. (3.31) must now be solved for the unknown surface components of the displacement and traction rates in terms of the prescribed ones. In order to do this, the plastic strain rate tensor in the volume integral must be determined from a suitable constitutive equation. This depends on the plasticity constitutive model that is chosen to describe material behavior. For example, for an isotropic strain hardening Von Mises material, this takes the form

$$\dot{\varepsilon}_{ij}^{(\mathrm{pl})}=\frac{3}{2}\frac{\dot{\varepsilon}^{(\mathrm{pl})}}{\sigma}s_{ij}=\frac{9}{4}\frac{s_{kl}\dot{\sigma}_{kl}s_{ij}}{h\sigma^2} \qquad (3.32)$$

where $$\sigma=\sqrt{\frac{3}{2}s_{ij}s_{ij}}$$

and $\dot{\varepsilon}^{(\mathrm{pl})}=\sqrt{\frac{2}{3}\dot{\varepsilon}_{ij}^{(\mathrm{pl})}\dot{\varepsilon}_{ij}^{(\mathrm{pl})}}$ are the stress and plastic strain rate invariants, respectively, and h is the plastic hardening modulus, the slope of the uniaxial stress–plastic strain (σ–$\varepsilon^{(\mathrm{pl})}$) curve. More general plasticity models, based on more realistic yield criteria, hardening laws and flow rules exist in the literature (e.g., references 8,10,11) but these will not be discussed further in this book. The dependence of the plastic strain rate on the stress rate typically requires iterations within a load step in a numerical solution procedure. Riccardella[4] has suggested a method for getting around this complication for planar problems.

The above transformation uses body forces based on plastic strain rates. An alternative formulation with body forces based on initial stresses is due to Banerjee *et al.*[12] In this case, eqn. (3.31) takes the form

$$c_{ij}(P)\dot{u}_i(P)=\int_{\partial B}\{U_{ij}(P,Q)\dot{\tau}_i(Q)-T_{ij}(P,Q)\dot{u}_i(Q)\}\mathrm{d}s_Q + \int_B B_{ikj}(p,q)\dot{\sigma}'_{ik}(q)\mathrm{d}v_q \qquad (3.33)$$

where B_{ijk} are the strains due to point loads in the Kelvin solution, i.e., from eqn. (3.8)

$$\varepsilon_{ij}^{*}=\frac{1}{2}(U_{ik,j}+U_{jk,i})e_k=B_{ijk}e_k \tag{3.34}$$

The explicit form of B_{ijk} is

$$B_{ijk}=-\frac{1}{16\pi G(1-\nu)r^2}\{(1-2\nu)(r_{,i}\delta_{jk}+r_{,j}\delta_{ki}) - r_{,k}\delta_{ij}+3r_{,i}r_{,j}r_{,k}\} \tag{3.35}$$

For plasticity problems with $\dot{\varepsilon}_{kk}^{(\mathrm{pl})}=0$, the initial stress tensor $\sigma'_{ik}=2G\dot{\varepsilon}_{ik}^{(\mathrm{pl})}$ and eqn. (3.33) is identical to (3.31).

Once the traction and displacement rates have been obtained over the entire surface ∂B, these can be used in eqn. (3.30) to determine the displacement rate field throughout the body. Finally, the stress rates must be calculated at any internal point p.

3.2.3 Stress Rates

As discussed before in the section on elasticity, several possible strategies can be used for the determination of strain rates from the displacement rates. The first is the numerical differentiation of the displacement rate field by finite difference or finite element techniques. The second is the analytical differentiation of eqn. (3.30) at an internal source point p. This must be carried out with care since direct differentiation of $U_{ij,k}$ under the integral sign leads to a nonintegrable singularity of the type r^{-3}.[13] A comparison of results from the two methods, for the expansion of an elasto-plastic cylinder in plane strain under increasing internal pressure, appears in a recent paper.[14] An alternative approach is the analytical evaluation of the volume integral over a typical volume element in which the integrand becomes singular, and then differentiation of this integral at the point p. This approach has been carried out successfully for planar viscoplastic problems.[15] Yet another possibility is the use of a 'strain-rate gradient' formulation. These methods will be discussed in some detail in the next section on viscoplasticity.

The stress rates at an internal point can be obtained from the strain rates by first using eqn. (3.21) to get the elastic strain rates and then Hooke's law in rate form. The stress rates at a point on the boundary can be obtained from an extended form of eqns. (3.18) to (3.20) in rate form. These equations, and some thoughts on numerical implementation of the integral equations, are presented in the next section where the analogous problem of time-dependent viscoplastic deformation is discussed.

3.3 TIME-DEPENDENT VISCOPLASTICITY

3.3.1 Governing Differential Equations

The governing differential equations in this case have exactly the same form as those for plasticity, with two important differences. The first is that the plastic strain rates $\dot{\varepsilon}_{ij}^{(\mathrm{pl})}$ in each of the plasticity equations are now replaced by nonelastic strain rates $\dot{\varepsilon}_{ij}^{(\mathrm{n})}$. The second is that a superposed dot now denotes a rate with respect to real time. The nonelastic strain rate tensor can either be the creep strain rate tensor $\dot{\varepsilon}_{ij}^{(\mathrm{c})}$ from eqn. (2.8) or $\dot{\varepsilon}_{ij}^{(\mathrm{n})}$ from a combined creep–plasticity constitutive model eqn. (2.11). Time and rate effects are now explicitly included in the constitutive model. Further, the nonelastic strain rates are now functions of the current values of the stress, temperature, and possibly state variables, but not of stress rates. In the interest of simplicity, thermal strains are ignored in this section, but can be included without difficulty.[16] Thus, for example, the kinematic equations have the form

$$\dot{\varepsilon}_{ij}=\dot{\varepsilon}_{ij}^{(\mathrm{e})}+\dot{\varepsilon}_{ij}^{(\mathrm{n})}=\frac{1}{2}(\dot{u}_{i,j}+\dot{u}_{j,i}) \tag{3.36}$$

in this discussion.

3.3.2 Boundary Element Formulation for Displacement Rates

Following the same chain of reasoning as for plasticity, the displacement rate equation for an internal point p can be written as (see eqn. 3.29)[15,16,17]

$$\begin{aligned}\dot{u}_j(p) &= \int_{\partial B}\{U_{ij}(p,Q)\dot{\tau}_i(Q)-T_{ij}(p,Q)\dot{u}_i(Q)\}\,\mathrm{d}s_Q\\ &\quad+\int_{\partial B}2GU_{ij}(p,Q)\dot{\varepsilon}_{ik}^{(\mathrm{n})}(Q)n_k(Q)\,\mathrm{d}s_Q\\ &\quad-\int_{B}2GU_{ij}(p,q)\dot{\varepsilon}_{ik,k}^{(\mathrm{n})}(q)\mathrm{d}v_q\end{aligned} \tag{3.37}$$

where explicit expressions for $\dot{\tau}_i'$ and $\dot{F}_i'$ have been substituted and physical body forces have been set to zero.

Using the divergence theorem on the volume integral, this equation can be written as

$$\dot{u}_j(p)=\int_{\partial B}\{U_{ij}(p,Q)\dot{t}_i(Q)-T_{ij}(p,Q)\dot{u}_i(Q)\}\,\mathrm{d}s_Q$$
$$+\int_B 2GU_{ij,k}(p,q)\dot{\varepsilon}_{ik}^{(n)}(q)\mathrm{d}v_q \tag{3.38}$$

which is analogous to eqn. (3.30). Finally, for a boundary point,

$$c_{ij}(P)\dot{u}_i(P)=\int_{\partial B}\{U_{ij}(P,Q)\dot{t}_i(Q)-T_{ij}(P,Q)\dot{u}_i(Q)\}\mathrm{d}s_Q$$
$$+\int_B 2GU_{ij,k}(P,q)\dot{\varepsilon}_{ik}^{(n)}(q)\mathrm{d}v_q \tag{3.39}$$

In this integral equation, the volume integral is known at any time through the constitutive model (eqn. (2.11)). Thus, the unknown quantities only appear on the boundary of a body. The *size of a problem*, in a numerical solution procedure, therefore depends only on the *boundary discretization* and internal discretization is required merely for the purpose of evaluating *integrals with known integrands*. This is generally carried out by Gaussian quadrature.

3.3.3 Stress Rates

As discussed before, several strategies are possible for the determination of strain rates at an internal point p.

Numerical differentiation. A finite difference or finite element approach can be used to obtain displacement rate gradients. In the latter approach the displacement rates can be approximated by suitable interpolation functions within each element and then differentiated piecewise. In this case, in general, discontinuities in displacement rate gradients will exist across inter-element boundaries.

Analytical differentiation. Analytical differentiation of the volume integral in eqn. (3.38) must be carried out with care. Equation (3.38), differentiated at a source point, can be written as

$$\dot{u}_{j,L}(p)=\int_{\partial B}\{U_{ij,L}(p,Q)\dot{t}_i(q)-T_{ij,L}(p,Q)\dot{u}_i(Q)\}\mathrm{d}s_Q$$
$$+\frac{\partial}{\partial x_L}\int_B 2GW_{jik}(p,q)\dot{\varepsilon}_{ik}^{(n)}(q)\mathrm{d}v_q \tag{3.40}$$

where $W_{ijk}=U_{ij,k}$

The kernel W_{jik} has a singularity of order r^{-2} and cannot be differentiated directly under the integral sign. One possibility is to evaluate the volume integral analytically for an arbitrary source point p and then differentiate this integral at p. This typically requires interpolation of the nonelastic strain rate field within a volume element, evaluation of the integral over an element in which the integrand is singular, and then differentiation of this integral at p. This has been successfully carried out for planar problems.[15]

Another alternative is the method suggested by Bui.[13] In this case a small sphere $B_\eta(p)$, of radius η, centered at p, is isolated from the volume B, and the volume integral is written as

$$\int_B 2GW_{jik}(p,q)\dot{\varepsilon}_{ik}^{(n)}(q)\mathrm{d}v_q = \int_{B-B_\eta(p)} 2GW_{jik}(p,q)\dot{\varepsilon}_{ik}^{(n)}(q)\mathrm{d}v_q$$

$$+\int_{B_\eta(p)} 2GW_{jik}(p,q)\dot{\varepsilon}_{ik}^{(n)}(q)\mathrm{d}v_q \qquad (3.41)$$

Convected differentiation of the integral over $B-B_\eta(p)$ using the formula

$$\frac{\partial}{\partial x_L}\int_{B-B_\eta(p)} I_j(p,q)\mathrm{d}v_q = \int_{B-B_\eta(p)} I_{j,L}(p,q)\mathrm{d}v_q$$

$$-\int_{\partial B_\eta} I_j(p,Q)n_l(Q)\mathrm{d}s_Q$$

and dropping the term $\dfrac{\partial}{\partial x_L}\displaystyle\int_{B_\eta(p)} 2GW_{jik}\dot{\varepsilon}_{ik}^{(n)}\mathrm{d}v$ (since it vanishes as $\eta\to 0$), leads to the equation[13]

$$\dot{u}_{j,L}(p) = \int_{\partial B}\{U_{ij,L}(p,Q)\dot{\tau}_i(Q) - T_{ij,L}(p,Q)\dot{u}_i(Q)\}\mathrm{d}s_Q$$

$$+\int_{B-B_\eta} 2GW_{jik,L}(p,q)\dot{\varepsilon}_{ik}^{(n)}(q)\mathrm{d}v_q$$

$$+\frac{8-10\nu}{15(1-\nu)}\dot{\varepsilon}_{jl}^{(n)}(p) \qquad (3.42)$$

The integral over $B-B_\eta$, in the above expression, yields the principal value of the singular integral as $\eta \to 0$. This integral, however, still has a kernel with a singularity of order r^{-3} and must be determined carefully.[18] The last term, from the integral over the surface of the sphere ∂B_{η}, is the convected term and must not be neglected.

Strain rate gradient formula. Use of this method avoids the kernel $W_{jik,L}$ which has a singularity of order r^{-3}. The equation (3.37) for this displacement rate field can be used instead of the equivalent expression from eqn. (3.38). This gives (see also reference 19, where a similar idea has been suggested independently by Telles and Brebbia)

$$\dot{u}_{j,L}(p) = \int_{\partial B} \{U_{ij,L}(p,Q)\dot{t}_i(Q) - T_{ij,L}(p,Q)\dot{u}_i(Q)\} \mathrm{d}s_Q$$
$$+ \int_{\partial B} 2GU_{ij,L}(p,Q)\dot{\varepsilon}_{ik}^{(n)}(Q) n_k(Q) \mathrm{d}s_Q$$
$$- \int_{B} 2GU_{ij,L}(p,q)\dot{\varepsilon}_{ik,k}^{(n)}(q) \mathrm{d}v_q \tag{3.43}$$

where there is no difficulty in differentiating any of the kernels under the integral sign. Of course, this method requires the accurate evaluation of the nonelastic strain rates over the boundary ∂B, and this, in turn, requires the stresses accurately on the boundary at any time. The boundary stresses and stress rates can be obtained from the equations given in the next section. Equation (3.43) has the advantage of having a kernel of singularity r^{-2} in the volume integral and is convenient to use in cases where the kernels cannot be easily integrated analytically over a typical volume element. This is the case if a volume element is irregular or for axisymmetric problems formulated with axisymmetric kernels, as is described in a later chapter. Further discussion of numerical integration of singular kernels is deferred to the subsection on numerical implementation later in this chapter and in subsequent chapters.

3.3.4 Boundary Stress Rates

The boundary stress rates can be obtained from an extended version of eqns. (3.18)–(3.20) in rate form. Thus,

$$\dot{\sigma}_{ij} = \lambda \dot{u}_{k,k}\delta_{ij} + G(\dot{u}_{i,j} + \dot{u}_{j,i}) - 2G\dot{\varepsilon}_{ij}^{(n)} \tag{3.44}$$

$$\dot{t}_i = \dot{\sigma}_{ij} n_j \tag{3.45}$$

$$\frac{\partial \dot{u}_i}{\partial s} = \dot{u}_{i,j} t_j \tag{3.46}$$

The method outlined in Section 3.1.3 for elastic problems can be used without difficulty since $\dot{\varepsilon}_{ij}^{(n)}$ is known at any time as a function of stresses and state variables through the constitutive eqn. (2.11).

3.3.5 An Analytical Example

The analytical example considered here is that of a sphere, of internal radius a and external radius b, subjected to internal and external pressure histories $p_i(t)$ and $p_o(t)$, respectively.[16] The sphere undergoes nonelastic deformation. The temperature field is uniform throughout the sphere.

The only nonzero component of displacement is the radial displacement u which is a function of the radial location R and time t. The nonzero components of stress and strain tensors, namely σ_{RR}, $\sigma_{\theta\theta}$ $(=\sigma_{\varphi\varphi})$, ε_{RR} and $\varepsilon_{\theta\theta}$ $(=\varepsilon_{\varphi\varphi})$ are functions of R and t only. The usual notation is used where R, θ and ϕ denote the radial and two mutually perpendicular orthogonal circumferential directions (Fig. 3.1).

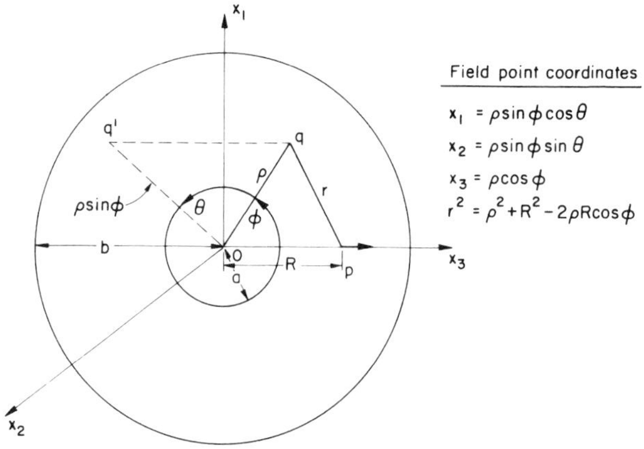

FIG. 3.1. Geometry of the sphere problem.

A unit load is applied in the x_3 direction at a point on the x_3 axis as shown in Fig. 3.1. The integral formulae (3.38) and (3.39) for the displacement rate, now assume the form (no sum on ρ or θ)

$$\kappa\dot{u}(R) = \int_{\substack{\text{sphere}\\ \rho=a}} (U_{\rho 3}\dot{\tau}_\rho - T_{\rho 3 -}\dot{u}_\rho)\mathrm{d}s$$

$$+ \int_{\substack{\text{sphere}\\ \rho=b}} (U_{\rho 3}\dot{\tau}_\rho - T_{\rho 3}\dot{u}_\rho)\mathrm{d}s$$

$$+2G\int_v \{W_{3\rho\rho}\dot{\varepsilon}^{(n)}_{\rho\rho}+(W_{3\theta\theta}+W_{3\varphi\varphi})\dot{\varepsilon}^{(n)}_{\theta\theta}\}\mathrm{d}v$$

$$=2\pi a^2\dot{p}_\mathrm{i}\int_0^\pi U_{\rho 3}(R;a,\phi)\sin\phi\,\mathrm{d}\phi$$

$$-2\pi b^2\dot{p}_\mathrm{o}\int_0^\pi U_{\rho 3}(R;b,\phi)\sin\phi\,\mathrm{d}\phi$$

$$+2\pi a^2\dot{u}(a)\int_0^\pi T_{\rho 3}(R;a,\phi)\sin\phi\,\mathrm{d}\phi$$

$$-2\pi b^2\dot{u}(b)\int_0^\pi T_{\rho 3}(R;b,\phi)\sin\phi\,\mathrm{d}\phi$$

$$+2\pi\int_0^b\int_0^\pi [W_{3\rho\rho}(R;\rho,\phi)\dot{\varepsilon}^{(n)}_{\rho\rho}(\rho)$$
$$+\{W_{3\theta\theta}(R;\rho,\phi)+W_{3\varphi\varphi}(R;\rho,\phi)\}\dot{\varepsilon}^{(n)}_{\theta\theta}(\rho)]\rho^2 \sin\phi\,\mathrm{d}\phi\,\mathrm{d}\rho \tag{3.47}$$

where $\kappa=1$ for an interior point and $\kappa=\frac{1}{2}$ for a boundary point, $T_{\rho 3-}=-T_{3\rho}$, $W_{ijk}=U_{ij,k}$ and the kernels, in terms of the distance r between the source point and the field point, are

$$U_{\rho 3}(R;\rho,\phi)=\frac{1}{16\pi(1-\nu)G}\left\{\frac{(3-4\nu)\cos\phi}{r}+\frac{(\rho-R\cos\phi)(\rho\cos\phi-R)}{r^3}\right\} \tag{3.48}$$

$$T_{\rho 3}(R;\rho,\phi)=-\frac{1}{8\pi(1-\nu)}\left[(1-2\nu)\left\{\frac{2(\rho-R\cos\phi)\cos\phi}{r^3}-\frac{(\rho\cos\phi-R)}{r^3}\right\}+\frac{3(\rho\cos\phi-R)(\rho-R\cos\phi)^2}{r^5}\right] \tag{3.49}$$

$$2GW_{3\rho\rho}(R;\rho,\phi)=-\frac{1}{8\pi(1-\nu)}\left\{\frac{2(1-2\nu)(\rho-R\cos\phi)\cos\phi}{r^3}-\frac{(\rho\cos\phi-R)}{r^3}+\frac{3(\rho\cos\phi-R)(\rho-R\cos\phi)^2}{r^3}\right\} \tag{3.50}$$

$$2GW_{3\theta\theta}(R;\rho,\phi)=\frac{\rho\cos\phi-R}{8\pi(1-\nu)r^3} \tag{3.51}$$

$$2GW_{3\varphi\varphi}(R;\rho,\phi)=-\frac{1}{8\pi(1-\nu)}\left\{\frac{-2(1-2\nu)R\sin^2\phi}{r^3}-\frac{(\rho\cos\phi-R)}{r^3}+\frac{3(\rho\cos\phi-R)R^2\sin^2\phi}{r^5}\right\} \tag{3.52}$$

with $r^2=\rho^2+R^2-2\rho R\cos\phi$. Using the equation $\dot{\varepsilon}_{RR}^{(n)}+2\dot{\varepsilon}_{\theta\theta}^{(n)}=0$ to guarantee incompressible nonelastic strains, the double integral in eqn. (3.47) reduces to the simple form

$$-\frac{(1-2\nu)R}{(1-\nu)}\int_R^b\frac{\dot{\varepsilon}_{RR}^{(n)}(\rho)\mathrm{d}\rho}{\rho}$$

The appropriate integrals are listed in Table 3.1.

In order to solve for the displacement rates, eqn. (3.47) is first used with $R=a$ and then with $R=b$. This leads to linear algebraic equations

TABLE 3.1

INTEGRALS FOR SPHERE PROBLEM

$f(\phi)$	$\int_0^\pi f(\phi)\sin\phi\,\mathrm{d}\phi$		
	$\rho>R$	$\rho=R$	$\rho<R$
$\cos\phi/r$	$\frac{2R}{3\rho^2}$	$\frac{2}{3R}$	$\frac{2\rho}{3R^2}$
$\frac{(\rho-R\cos\phi)(\rho\cos\phi-R)}{r^3}$	$-\frac{2R}{3\rho^2}$	$-\frac{2}{3R}$	$-\frac{2\rho}{3R^2}$
$\frac{2\cos\phi(\rho-R\cos\phi)}{r^3}$	$\frac{8R}{3\rho^3}$	$\frac{2}{3R^2}$	$-\frac{4}{3R^2}$
$\frac{3(\rho\cos\phi-R)(\rho-R\cos\phi)^2}{r^5}$	0	$-\frac{1}{R^2}$	$-\frac{2}{R^2}$
$\frac{(\rho\cos\phi-R)}{r^3}$	0	$-\frac{1}{R^2}$	$-\frac{2}{R^2}$
$\frac{2R\sin^2\phi}{r^3}$	$\frac{8R}{3\rho^3}$	$\frac{8}{3R^2}$	$\frac{8}{3R^2}$
$\frac{3(\rho\cos\phi-R)R^2\sin^2\phi}{r^5}$	0	$-\frac{2}{R^2}$	$-\frac{4}{R^2}$

$r^2=\rho^2+R^2-2\rho R\cos\phi$

for the boundary displacement rates $\dot{u}(a)$ and $\dot{u}(b)$. Solving these,

$$\begin{aligned}\dot{u}(a)=&\frac{\dot{p}_{\mathrm{i}}a^3}{(b^3-a^3)E}\left\{a(1-2\nu)+\frac{b^3}{2a^2}(1+\nu)\right\}\\&-\frac{\dot{p}_{\mathrm{o}}b^3}{(b^3-a^3)E}\left\{a(1-2\nu)+\frac{a}{2}(1+\nu)\right\}\\&-\frac{3ab^3}{2(b^3-a^3)}\int_a^b\frac{\dot{\varepsilon}^{(\mathrm{n})}_{RR}(\rho)\mathrm{d}\rho}{\rho}\end{aligned}\tag{3.53}$$

$$\begin{aligned}\dot{u}(b)=&\frac{\dot{p}_{\mathrm{i}}a^3}{(b^3-a^3)E}\left\{b(1-2\nu)+\frac{b(1+\nu)}{2}\right\}\\&-\frac{\dot{p}_{\mathrm{o}}b^3}{(b^3-a^3)E}\left\{b(1-2\nu)+\frac{a^3(1+\nu)}{2b^2}\right\}\\&-\frac{3a^3b}{2(b^3-a^3)}\int_a^b\frac{\dot{\varepsilon}^{(\mathrm{n})}_{RR}(\rho)\mathrm{d}\rho}{\rho}\end{aligned}\tag{3.54}$$

Finally, eqn. (3.47) is used at an internal point to give the displacement rate equation

$$\begin{aligned}\dot{u}(R)=&\frac{\dot{p}_{\mathrm{i}}a^3}{(b^3-a^3)E}\left\{(1-2\nu)R+\frac{(1+\nu)b^3}{2R^2}\right\}\\&-\frac{\dot{p}_{\mathrm{o}}b^3}{(b^3-a^3)E}\left\{(1-2\nu)R+\frac{(1+\nu)a^3}{2R^2}\right\}\\&+\frac{1}{1-\nu}\left[(1-2\nu)R\int_a^R\frac{\dot{\varepsilon}^{(\mathrm{n})}_{RR}\mathrm{d}\rho}{\rho}\right.\\&\left.-\frac{b^3}{b^3-a^3}\left\{(1-2\nu)R+\frac{(1+\nu)a^3}{2R^3}\right\}\int_a^b\frac{\dot{\varepsilon}^{(\mathrm{n})}_{RR}\mathrm{d}\rho}{\rho}\right]\end{aligned}\tag{3.55}$$

The stress rates can be obtained from the displacement rates through the kinematic equations and Hooke's law

$$\dot{\varepsilon}_{RR}=\frac{\partial\dot{u}}{\partial R},\ \dot{\varepsilon}_{\theta\theta}=\frac{\dot{u}}{R}\tag{3.56}$$

$$\dot{\sigma}_{RR}=\frac{(1-\nu)E(\dot{\varepsilon}_{RR}-\dot{\varepsilon}^{(\mathrm{n})}_{RR})}{(1-2\nu)(1+\nu)}+\frac{2\nu E(\dot{\varepsilon}_{\theta\theta}-\dot{\varepsilon}^{(\mathrm{n})}_{\theta\theta})}{(1-2\nu)(1+\nu)}\tag{3.57}$$

$$\dot{\sigma}_{\theta\theta}=\frac{\nu E(\dot{\varepsilon}_{RR}-\dot{\varepsilon}^{(\mathrm{n})}_{RR})}{(1-2\nu)(1+\nu)}+\frac{E(\dot{\varepsilon}_{\theta\theta}-\dot{\varepsilon}^{(\mathrm{n})}_{\theta\theta})}{(1-2\nu)(1+\nu)}\tag{3.58}$$

to give

$$\dot{\sigma}_{RR}=\left(\frac{E}{1-v}\right)\left(\int_a^R \frac{\dot{\varepsilon}_{RR}^{(n)}}{\eta}\mathrm{d}\eta-\frac{(R^3-a^3)\,b^3}{(b^3-a^3)\,R^3}\int_a^b \frac{\dot{\varepsilon}_{RR}^{(n)}}{\eta}\mathrm{d}\eta\right)$$
$$-\dot{p}_{\mathrm{i}}\frac{(b^3-R^3)\,a^3}{(b^3-a^3)\,R^3}-\dot{p}_{\mathrm{o}}\frac{(R^3-a^3)\,b^3}{(b^3-a^3)\,R^3} \tag{3.59}$$

$$\dot{\sigma}_{\theta\theta}=\left(\frac{E}{1-v}\right)\left(\int_a^R \frac{\dot{\varepsilon}_{RR}^{(n)}}{\eta}\mathrm{d}\eta-\frac{(2R^3+a^3)\,b^3}{2(b^3-a^3)\,R^3}\int_a^b \frac{\dot{\varepsilon}_{RR}^{(n)}}{\eta}\mathrm{d}\eta\right.$$
$$\left.+\frac{\dot{\varepsilon}_{RR}^{(n)}}{2}\right)+\dot{p}_{\mathrm{i}}\frac{(b^3+2R^3)\;a^3}{(b^3-a^3)\;2R^3}-\dot{p}_{\mathrm{o}}\frac{(2R^3+a^3)\;b^3}{(b^3-a^3)\;2R^3} \tag{3.60}$$

These equations check out against those derived directly from the governing differential equations in spherical coordinates.[20]

3.3.6 Numerical Implementation

Boundary value problems for bodies with irregular geometry must, in general, be solved by numerical methods. Some remarks on the numerical implementation of the integral equations are made in this section. Further details are given in subsequent chapters where numerical solutions for specific problems are presented and discussed.

The first step is the discretization of the three-dimensional body into surface elements and internal cells. As mentioned before, the internal discretization is necessary for the evaluation of volume integrals with known integrands. A discretized version of the boundary integral eqn. (3.39) for the displacement rate can be written as

$$c_{ij}(P_M)\dot{u}_i(P_M)=\sum_{N_s}\int_{\Delta s_N} U_{ij}(P_M,Q)\dot{\tau}_i(Q)\mathrm{d}s_Q$$
$$-\sum_{N_s}\int_{\Delta s_N} T_{ij}(P_M,Q)\dot{u}_i(Q)\mathrm{d}s_Q$$
$$+\sum_{n_i}\int_{\Delta v_n} 2GW_{ijk}(P_M,q)\dot{\varepsilon}_{ik}^{(n)}(q)\mathrm{d}v_q \tag{3.61}$$

where the surface of the body ∂B has been divided into N_s boundary elements and the interior into n_i internal cells and $\dot{u}_i(P_M)$ are the components of the displacement rates at a point P which coincides with node M.

Suitable shape functions must now be chosen for the variation of traction and displacement rates on the surface elements Δs_N and the variation of the nonelastic strain rates over an internal cell Δv_n. *Special care must be taken for integration over an element in which a kernel is singular*, i.e. over an element which contains the source point P_M. If the elements have simple geometry and the shape functions have simple form, it might be possible to carry out integrals of kernels analytically. Cruse,[3] for example, has solved three-dimensional *elasticity* problems with flat triangular surface elements and piecewise uniform tractions and displacements on these elements. Each surface element is denoted by its centroidal point so that $c_{ij}=(\frac{1}{2})\delta_{ij}$. In this special case, it is possible to evaluate the integrals

$$\Delta U_{ij}=\int_{\Delta s_N} U_{ij}(P_M, Q)\mathrm{d}s_Q$$

$$\Delta T_{ij}=\int_{\Delta s_N} T_{ij}(P_M, Q)\mathrm{d}s_Q$$

exactly, knowing the size, orientation and location of the surface element Δs_N and the point P_M.

Analytical integration of this type is not possible in general. Integration of these kernels over elements which are far from the source point can generally be carried out conveniently and very accurately by Gaussian integration. Integration in singular and near-singular situations, however, is another matter and special methods are often useful in such cases. One such method is the use of rigid body displacements in the absence of creep. As mentioned before, this gives a stress-free state and leads to the auxiliary equation (3.14). This equation, written in the form[21]

$$c_{ij}(P)+\int_{\partial B_c} T_{ij}(P,Q)\mathrm{d}s_Q$$

$$=-\int_{\partial \hat{B}} T_{ij}(P,Q)\mathrm{d}s_Q \qquad (3.62)$$

(where ∂B_c is the part of ∂B which contains P and $\partial \hat{B}$ is the rest) is very useful since it relates the singular integrals over ∂B_c (together with the tensor c_{ij}) with regular integrals over the rest of the surface. The regular integrals over $\partial \hat{B}$ can be evaluated accurately by Gaussian quadrature.

Another comment pertains to the numerical modelling of possible jumps in normals or prescribed tractions across boundaries of surface elements. A simple way around this is to place source points inside rather than on the boundaries of surface elements. This is easily done in the boundary element method since the source points are really sampling points (or boundary collocation points). These points, for example, need not lie on the vertices of a triangular surface element over which traction and displacement rates are assumed to vary in a linear fashion. An alternative is to specify, if necessary, a 'zero length' element between coincident source points and assign different values of physical quantities on these points. This latter approach has been used for the solution of planar and axisymmetric viscoplastic problems which are described later in this book.

Numerical solutions for three-dimensional viscoplasticity problems do not exist at this time, so comments about the volume integral in equation (3.61) are *speculative in nature.* Analytical integration of W_{ijk}, if possible, is no doubt the best approach. If this is not possible, a transformation is suggested here for the evaluation of the integral over an element in which the kernel is singular. This transformation has proved to be very useful for axisymmetric problems.

The volume element in question is assumed to be tetrahedral with the point P_M lying on one of its vertices. Using local spherical–polar coordinates centered at P_M, the integral W_{ijk} over such an element has the form

$$\int_{\Delta v_n} W_{ijk}(P_M, q)\mathrm{d}v_q = \int_{\text{suitable limits}} \frac{f(r,\theta,\phi)}{r^2} r^2 \sin\phi \mathrm{d}r\mathrm{d}\theta\mathrm{d}\phi \tag{3.63}$$

One option is to integrate f/r^2 by Gaussian quadrature. A better approach appears to be the numerical evaluation of the integral

$$\int f(r,\theta,\phi)\sin\phi \mathrm{d}r\mathrm{d}\theta\mathrm{d}\phi$$

over a suitably transformed domain by Gaussian quadrature. The latter approach has the advantage of using a regular integrand in the integral in question.

The consequence of numerical discretization is to reduce eqn. (3.61) into an algebraic form of the type

$$[A]\{\dot{u}\}+[B]\{\dot{\tau}\}=\{\dot{b}\} \tag{3.64}$$

where the coefficient matrices $[A]$ and $[B]$ contain integrals of the kernels and the shape functions and the vector $\{\dot{b}\}$ contains the non-elastic strain rates and the kernel W_{ijk}. The vector $\{\dot{b}\}$ is known at any time through the constitutive eqn. (2.11) and the size of the matrices [A] and [B] depend only on the boundary discretization. A simple switching of the columns of [A] and [B] is carried out in order to determine the unknown components of the traction and displacement rates in terms of the prescribed ones. The displacement rate field throughout the body can be obtained from eqn. (3.38) discretized in an analogous manner.

The stress rates on the boundary and inside the body must now be determined from the displacement rates. The boundary stress rates can be determined easily from the algorithm set forth in Section 3.3.4. It is felt that, for internal stress rates, it is advantageous to use a pointwise algorithm rather than an elementwise approach. The former method retains one of the important features of the boundary element method in that discontinuities in stresses or stress rates across boundaries of internal cells do not occur in this case. This is particularly beneficial in improving the accuracy of numerical solutions of these viscoplastic problems where the nonelastic strain rates are typically functions of high powers of the stress components. If a pointwise algorithm is followed, it appears easiest to use a discretized version of eqn. (3.43). This takes the form

$$\begin{aligned}\dot{u}_{j,L}(p_m)=&\sum_{N_s}\int_{\Delta s_N} U_{ij,L}(p_m,Q)\dot{\tau}_i(Q)\mathrm{d}s_Q\\&-\sum_{N_s}\int_{\Delta s_N} T_{ij,L}(p_m,Q)\dot{u}_i(Q)\mathrm{d}s_Q\\&+\sum_{N_s}\int_{\Delta s_N} 2GU_{ij,L}(p_m,Q)\dot{\varepsilon}_{ik}^{(n)}(Q)n_k(Q)\mathrm{d}s_Q\\&-\sum_{n_i}\int_{\Delta v_n} 2GU_{ij,L}(p_m,q)\dot{\varepsilon}_{ik,k}^{(n)}(q)\mathrm{d}v_q\end{aligned} \qquad (3.65)$$

This step involves evaluation of integrals with known integrands at each time. The volume integral in eqn. (3.65), which contains terms of the type r^{-2}, can be evaluated through the transformation described earlier in this section. The nonelastic strain rates, of course, must be interpolated at least linearly over a volume element.

An alternative is the use of eqn. (3.42) where *special care* must be taken in the determination of the Cauchy principal value of the volume integral on elements where it becomes singular. This is necessary in view of the *strongly singular nature of the kernel* (r^{-3}) *in this case*. An analytical approach for two-dimensional problems using eqn. (3.42) has been suggested by Telles and Brebbia.[18] Numerical or analytical methods for three-dimensional problems have not yet been attempted.

A discussion of the complete solution procedure using these discretized equations is postponed to the next chapter.

REFERENCES

1. SOKOLNIKOFF, I. S. *Mathematical Theory of Elasticity*, McGraw-Hill, New York (1956).
2. RIZZO, F. J. An integral equation approach to boundary value problems of classical elastostatics, *Quarterly of Applied Mathematics*, **25**, 83–95 (1967).
3. CRUSE, T. A. Numerical solutions in three-dimensional elastostatics. *International Journal of Solids and Structures*, **5**, 1259–1274 (1969).
4. RICCARDELLA, P. C. *An Implementation of the Boundary-Integral Technique for Planar Problems of Elasticity and Elasto-Plasticity*. Report No. SM-73-10, Department of Mechanical Engineering, Carnegie Mellon University, Pittsburg, PA (1973).
5. CRUSE, T. A. An improved boundary-integral equation method for three-dimensional elastic stress analysis. *Computers and Structures*, **4**, 741–754 (1974).
6. LACHAT, J. C. *A Further Development of the Boundary-Integral Technique for Elastostatics*. Dissertation, University of Southampton, England (1975).
7. RIZZO, F. J. and SHIPPY, D. J. A formulation and solution procedure for the general nonhomogeneous elastic inclusion problem. *International Journal of Solids and Structures*, **4**, 1161–1179 (1968).
8. SWEDLOW, J. L. and CRUSE, T. A. Formulation of boundary integral equations for three-dimensional elasto-plastic flow. *International Journal of Solids and Structures*, **7**, 1673–1681 (1971).
9. MENDELSON, A. *Boundary-Integral Methods in Elasticity and Plasticity*. NASA Report No. TND-7418 (1973).
10. MENDELSON, A. *Plasticity—Theory and Applications*, Macmillan, New York (1970).
11. FUNG, Y. C. *Foundations of Solid Mechanics*, Prentice Hall, New York (1965).
12. BANERJEE, P. K., CATHIE, D. N. and DAVIES, T. G. Two and three-dimensional problems of elasto-plasticity. Chapter 4 in *Developments in Boundary Element Methods—1*, P. K. Banerjee and R. Butterfield (eds.), Applied Science Publishers Ltd, Barking, Essex, UK, 65–95 (1979).
13. BUI, H. D. Some remarks about the formulation of three-dimensional thermoelastoplastic problems by integral equations. *International Journal of Solids and Structures*, **14**, 935–939 (1978).
14. TELLES, J. C. F. and BREBBIA, C. A. New developments in elasto-plastic analysis, *Boundary Element Methods—Proceedings of the Third International Seminar*, Irvine, California, C. A. Brebbia (ed.), Springer-Verlag, Berlin-Heidelberg, 350–370 (1981).
15. MUKHERJEE, S. and KUMAR, V. Numerical analysis of time-dependent

inelastic deformation in metallic media using the boundary-integral equation method. American Society of Mechanical Engineers, *Journal of Applied Mechanics*, **45**, 785–790 (1978).
16. KUMAR, V. and MUKHERJEE, S. A boundary-integral equation formulation for time-dependent inelastic deformation in metals. *International Journal of Mechanical Sciences*, **19**, 713–724 (1977).
17. CHAUDONNERET, M. Methode des equations integrales appliquées ā la resolution de problèmes de viscoplasticité, *Journal de Mécanique Appliquée*, **1**, 113–132 (1977).
18. TELLES, J. C. F. and BREBBIA, C. A. The boundary element method in plasticity, *New Developments in Boundary Element Methods*, C. A. Brebbia (ed.), CML Publications, Southampton, UK, 295–317 (1980).
19. TELLES, J. C. F. and BREBBIA, C. A. Elasto-plastic boundary element analysis, *Nonlinear Finite Element Analysis in Structural Mechanics*, W. Wunderlich, E. Stein and K. J. Bathe (eds.), Springer-Verlag, Berlin, 403–433 (1981).
20. MUKHERJEE, S., KUMAR, V. and CHANG, K. J. Elevated temperature inelastic analysis of metallic media under time-varying loads using state variable theories, *International Journal of Solids and Structures*, **14**, 663–679 (1978).
21. RIZZO, F. J. and SHIPPY, D. J. An advanced boundary integral equation method for three-dimensional thermoelasticity, *International Journal for Numerical Methods in Engineering*, **11**, 1753–1768 (1977).

CHAPTER 4

Solution Strategy and Time Integration

A complete solution strategy, which can be used to solve a three-dimensional viscoplastic boundary value problem, is presented in this section. This strategy has been used to obtain numerical solutions for planar, axisymmetric, plate bending and torsion problems. These numerical solutions are presented in subsequent chapters.

4.1 SOLUTION STRATEGY

The solution strategy for three-dimensional viscoplastic boundary value problems is best summarized in Fig. 4.1. The initial values of the nonelastic strain components are taken to be zero and the initial values of the state variables, if any, are prescribed. The initial values of the stress and strain components are obtained from the solution of the corresponding elastic problem, i.e., from the elastic solution for an identical body with boundary conditions same as those prescribed for the original nonelastic problem at zero time. This elastic solution is obtained by the boundary element method by solving the appropriate integral equations from Sections 3.1.2 and 3.1.3.

The rates of the nonelastic strains at zero time are next obtained from the constitutive equations (e.g. eqn. (2.11)). These rates are used in eqn. (3.64) which is solved for the unspecified components of the boundary traction and displacement rates at $t=0$. The displacement rates throughout the body are next obtained from a discretized version of eqn. (3.37) or (3.38). Displacement rate gradients are then determined from a differentiated version of the displacement rate equation (like eqn. (3.42) or (3.43)), or by direct numerical differentiation of the displacement rate field. The stress rate components throughout the body follow from the

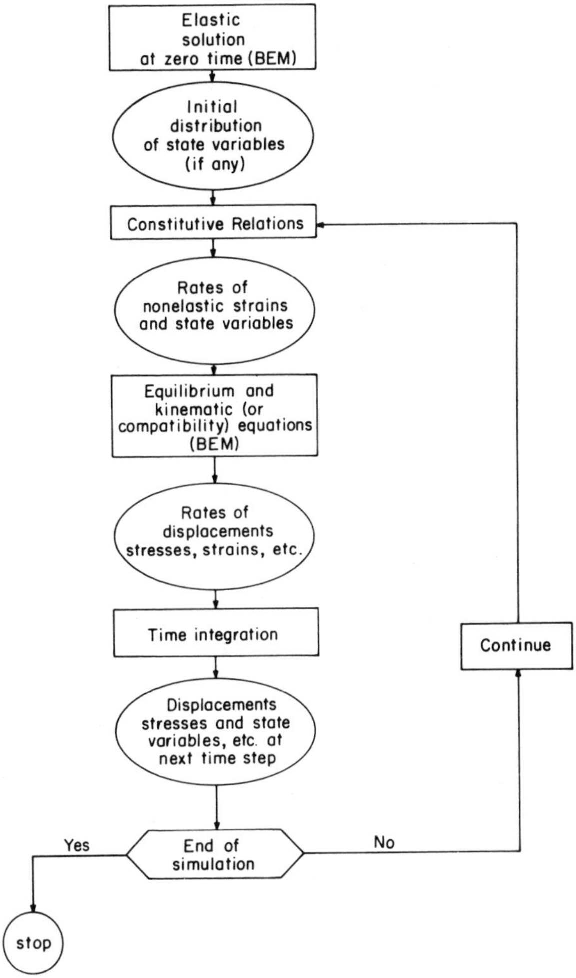

FIG. 4.1. Solution strategy.

components of the displacement rate gradients through Hooke's law. Finally, the rates of the state variables, if any, are determined from eqn. (2.12). It should be pointed out that the form of the constitutive equations presented in Chapter 2 makes this solution strategy particularly attractive, since the nonelastic strain rates are functions of the stress components and state variables, but not of stress rates as in work hardening plasticity (e.g. eqn. (3.32)).

These rates must now be used to find the values of the variables at a new time Δt, and so on, in order to obtain the time histories of the variables of interest. The constitutive equations for nonelastic deformation of metals generally lead to a stiff system of ordinary differential equations. Thus, the success of this solution method depends crucially on the strategy used for step-wise integration in time. Some possible strategies for time integration appear in the next section.

4.2 TIME INTEGRATION AND TIME-STEP SELECTION

It is very important, for these problems, to select an efficient time integration strategy with a suitable algorithm for time-step control. Many such strategies exist in the literature. A well-known method, designed for stiff systems of differential equations, is that due to Gear.[1-3] Another useful approach is due to Cormeau.[4] Some of these strategies include stability analysis and others do not.

A particular approach, which has worked very well for the numerical examples discussed in subsequent chapters of this book, is presented in this section.[5] The intention here is not to claim that this is the 'only' or 'best' possible approach, but simply to present one method which has proved successful for the solution of a variety of inelastic boundary value problems. The idea here is a combination of a classical method, such as one due to Euler or Adam, with a strategy for automatic time-step control at each step of the time integration process. This method is first discussed in the context of a single differential equation

$$\frac{dy}{dt}=F(y,t) \tag{4.1}$$

and generalizations to systems of equations are presented later in this section.

Automatic time-step control is based on comparison of a suitably defined error e with prescribed error limits e_{min} and e_{max}. The time step at the kth step, Δt_k, is defined on the basis of its estimate $\tilde{\Delta t}_k$, according as

$$\begin{aligned}
&e_{max}<e\text{: replace } \tilde{\Delta t}_k \text{ by } \tilde{\Delta t}_k/2\text{; recompute } e\\
&e\leq e_{max}\text{: } \Delta t_k=\tilde{\Delta t}_k\text{; compute } y_{k+1}\\
&e_{min}<e\leq e_{max}\text{: } \tilde{\Delta t}_{k+1}=\Delta t_k\\
&e\leq e_{min}\text{: } \tilde{\Delta t}_{k+1}=2\Delta t_k
\end{aligned}$$

The initial time step must be prescribed.

Some of the strategies described next are obtained from an expansion of y_{k+1} in terms of y_k:[6]

$$y_{k+1}=y_k+\Delta t_k(F_k+\tfrac{1}{2}\nabla F_k+\tfrac{5}{12}\nabla^2 F_k+\ \dots) \tag{4.2}$$

where $\nabla F_k=F_k-F_{k-1}$ is the first backward difference of F_k.

4.2.1 One-step Euler's Method

This is the simplest of time-integration schemes but, without automatic time-step control, is useless for most constitutive equations. Useful results have been obtained, however, by using the time-step control strategy described below.

The value of y_{k+1} in terms of y_k is

$$y_{k+1}=y_k+\Delta t_k F_k \tag{4.3}$$

and the error at this step used for automatic time-step control is

$$e=|\Delta t_k \nabla F_k|/|y_k| \tag{4.4}$$

4.2.2 Two-step Adam's Method

In this case

$$y_{k+1}=y_k+\Delta t_k F_k+(\Delta t_k/2)\nabla F_k \tag{4.5}$$

and

$$e=|\Delta t_k \nabla^2 F_k|/|y_k| \tag{4.6}$$

where $\nabla^2 F_k=F_k-2F_{k-1}+F_{k-2}$ is the second backward difference of F_k.

4.2.3 Predictor–Corrector Method

In this case a predictor is defined as

$$y^{\text{p}}_{k+1}=y_k+\Delta t_k F_k \tag{4.7}$$

and a corrector as

$$y^{\text{c}}_{k+1}=y_k+\Delta t_k F^{\text{p}}_{k+1}-\Delta t_k/2(F^{\text{p}}_{k+1}-F_k) \tag{4.8}$$

where

$$F^{\text{p}}_{k+1}=F(y^{\text{p}}_{k+1},\ t_{k+1})$$

Finally

$$e=\frac{|y^{\text{c}}_{k+1}-y^{\text{p}}_{k+1}|}{|y^{\text{c}}_{k+1}|}=\frac{|\Delta t_k \nabla F^{\text{p}}_{k+1}|}{2|y^{\text{c}}_{k+1}|} \tag{4.9}$$

where $\nabla F^{\mathrm{p}}_{k+1} = F^{\mathrm{p}}_{k+1} - F_k$ and the accepted value

$$y_{k+1} = (4/5)y^{\mathrm{c}}_{k+1} + (1/5)y^{\mathrm{p}}_{k+1} \tag{4.10}$$

4.2.4 Higher Order Predictor–Corrector Method

In this case

$$y^{\mathrm{p}}_{k+1} = y_k + \Delta t_k F_k + \Delta t_k/2(F_k - F_{k-1}) \tag{4.11}$$

$$y^{\mathrm{c}}_{k+1} = y_k + \Delta t_k F^{\mathrm{p}}_{k+1} - \Delta t_k/2(F^{\mathrm{p}}_{k+1} - F_k) \tag{4.12}$$

and

$$e = \frac{|y^{\mathrm{c}}_{k+1} - y^{\mathrm{p}}_{k+1}|}{|y^{\mathrm{c}}_{k+1}|} = \frac{|\Delta t_k \nabla^2 F_{k+1}|}{2|y^{\mathrm{c}}_{k+1}|} \tag{4.13}$$

The accepted value of y_{k+1} is obtained from eqn. (4.10).

4.2.5 Generalizations for Systems of Equations

This method can be easily extended for the case of systems of equations. The equations in this case can be written in the general form

$$\mathrm{d}y^{(\mathrm{i})}/\mathrm{d}t = F^{(\mathrm{i})}(y^{(i)}, t) \tag{4.14}$$

and, proceeding as before, an error vector $e^{(\mathrm{i})}$ is obtained. Now a suitable norm

$$e = L^{\mathrm{n}}(e^{(\mathrm{i})}) \tag{4.15}$$

is defined, and this norm is compared with $e_{\min}$ and $e_{\max}$. Three common norms that can be used are

$$L^{\infty} = \max|e^{(\mathrm{i})}|$$

$$L^1 = \sum_i |e^{(\mathrm{i})}|$$

$$L_2 = \sqrt{\sum_i (e^{(\mathrm{i})})^2}$$

Extension to multiaxial problems where the dependent variables are functions of both space and time is also straightforward. In this case each dependent variable is discretized at a finite number of space points (or elements) and a much larger number of ordinary differential equations results. This system of ordinary differential equations is now treated as discussed above.

Comparison of these strategies, for several typical uniaxial problems using Hart's constitutive equations, is given below. Gear's programs STIFF-0, STIFF-1 and STIFF-2, described in references 1–3, are also included in this comparison. STIFF-0 is a higher order predictor–corrector method, and STIFF-1 and STIFF-2 are packages for stiff systems, the former using an analytical and the latter a numerical representation of the Jacobian matrix.

4.3 COMPARISON OF VARIOUS STRATEGIES FOR UNIAXIAL PROBLEMS

Hart's equations have been integrated using the various integration strategies discussed in the previous section. Several input histories of stress and strain are considered and the results are compared in terms of accuracy and computational efficiency.

The material parameters used for 304 stainless steel at 400 °C are[7]

$$\lambda = 0{\cdot}15 \quad M = 7{\cdot}8 \quad m = 5$$
$$\mathcal{M} = 0{\cdot}132 \times 10^8 \text{ psi}$$
$$E = 0{\cdot}244 \times 10^8 \text{ psi} \quad \nu = 0{\cdot}298$$
$$\dot{\varepsilon}_0 = 3{\cdot}15 \text{ sec}^{-1} \text{ at } \sigma_0 = 10 \text{ ksi}$$
$$\dot{\varepsilon}^*_{sT} = 1{\cdot}269 \times 10^{-24} \text{ sec}^{-1} \text{ at } \sigma^*_s = 10 \text{ ksi}, \; T_B = 673 \text{ K}$$
$$\beta = 0{\cdot}179 \times 10^6 \text{ psi} \quad \delta = 1{\cdot}33$$

The parameters at 500°C are available in reference 7. The numerical results are shown in Figs. 4.2 and 4.3. Figure 4.2 shows results for prescribed histories of stress and Fig. 4.3 shows results for prescribed histories of strain. The error parameters used in these calculations are $e_{max} = 10^{-3}$ and $e_{min} = 10^{-4}$ and the initial time steps are 10^{-13} and 10^{-3} seconds for Figs. 4.2 and 4.3 respectively. All the methods give the same results within plotting accuracy.

Comparisons of the strategies with regard to number of function evaluations, number of time steps and computer time are shown in Tables 4.1–4.4. Combined methods like 4 and 3 on Tables 4.1–4.4 signify the use of the two-step Adam's method in the region outside the viscoplastic limit and the one-step Euler's method inside the viscoplastic limit. Gear's packages are in double precision and the rest are in single precision. It is seen from the tables that the simple Euler method and the 4–3 combination strategy work very well. It should be emphasized that a one-step method is not necessarily the best one. For example, the two-step Adam's method works better than the one-step Euler's method

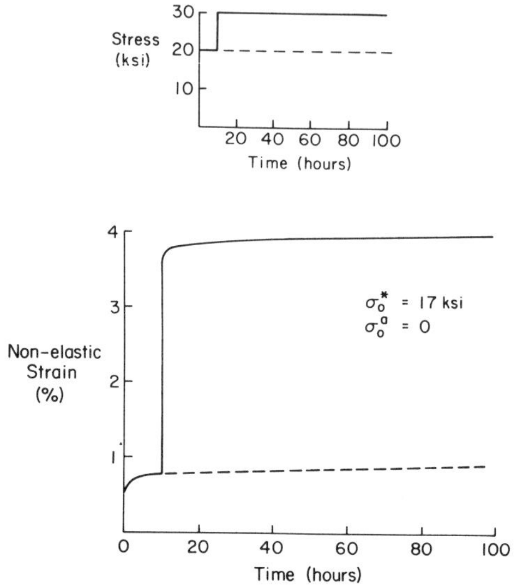

FIG. 4.2. Results for an annealed 304 SS bar at 400° C. Prescribed stress histories.

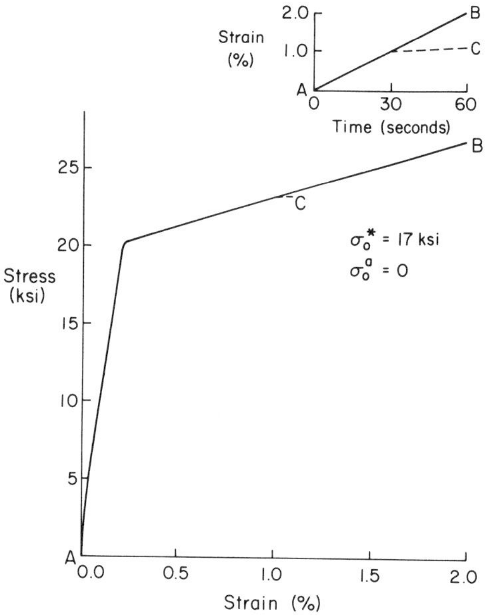

FIG. 4.3. Results for an annealed 304 SS bar. Prescribed strain histories.

TABLE 4.1

COMPARISON OF VARIOUS METHODS FOR CONSTANT STRESS CREEP OF ANNEALED 304 SS AT 400°C UP TO 100 HR.

Method	*Number of function evaluations*	*Number of time steps*	*CPU Time (sec) on IBM 370/168*
1. Predictor–corrector	1 555	741	0·654
2. Higher order predictor–corrector	1 488	684	0·573
3. One-step Euler	714	713	0·355
4. Two-step Adam	1 052	1 051	0·478
5. 2 and 1	1 376	646	0·492
6. 4 and 3	501	500	0·233
7. STIFF-0	6 396	1 644	3·73
8. STIFF-1	621(65 INV)	216	1·21
9. STIFF-2	546(29 INV)	186	0·59

$\sigma = 20$ ksi $\sigma_0^* = 17$ ksi

TABLE 4.2

COMPARISON OF VARIOUS METHODS FOR STRESS CHANGE TEST OF ANNEALED 304 SS AT 400°C

Method	*Number of function evaluations*	*Number of time steps*	*CPU Time (sec) on IBM 370/168*
1. Predictor–corrector	2 692	1 281	0·981
2. Higher order predictor–corrector	2 626	1 206	0·931
3. One-step Euler	1 176	1 175	0·530
4. Two-step Adam	1 899	1 898	0·845
5. 2 and 1	2 468	1 164	0·921
6. 4 and 3	1 003	1 002	0·440
7. STIFF-0	12 233	3 126	6·85
8. STIFF-1	1 386(96 INV)	418	2·81
9. STIFF-2	939(55 INV)	315	0·80

$\sigma = 20$ ksi for 0–10 hr.
$\sigma = 30$ ksi for 10–100 hr.
$\sigma_0^* = 17$ ksi

TABLE 4.3

COMPARISON OF VARIOUS METHODS FOR CONSTANT STRAIN RATE TENSION TEST OF ANNEALED 304 SS AT 400°C UP TO 2% STRAIN

Method	*Number of function evaluations*	*Number of time steps*	*CPU Time (sec) on IBM 370/168*
1. Predictor–corrector	1 505	743	0·552
2. Higher order predictor–corrector	1 624	700	0·669
3. One-step Euler	822	821	0·462
4. Two-step Adam	1 253	1 252	0·712
5. 2 and 1	1 442	647	0·538
6. 4 and 3	623	622	0·319

$\dot{\varepsilon}=0{\cdot}02$/min. $\sigma_0^*=17$ ksi

TABLE 4.4

COMPARISON OF VARIOUS METHODS FOR STRAIN RATE CHANGE TEST OF ANNEALED 304 SS AT 400° C

Method	*Number of function evaluations*	*Number of time steps*	*CPU Time (sec) on IBM 370/168*
1. Predictor–corrector	1 155	567	0·469
2. Higher order predictor–corrector	1 173	495	0·466
3. One-step Euler	693	692	0·396
4. Two-step Adam	813	812	0·459
5. 2 and 1	1 096	474	0·422
6. 4 and 3	527	526	0·283

$\dot{\varepsilon}=0{\cdot}02$/min for 0–30 sec.
$\dot{\varepsilon}=0{\cdot}002$/min for 30–60 sec.
$\sigma_0^*=17$ ksi

outside the viscoplastic limit and this results in the combined 4 and 3 strategy being better than the Euler's method used by itself (strategy 3). Adam's method is less efficient inside the viscoplastic limit because of oscillations caused by switching from eqn. (2.22) to (2.26) upon entering this region. Ongoing research results indicate that these oscillations can be reduced by a slight modification in the criterion for entering the viscoplastic limit.

Gear's STIFF-2 package works very well for the simulations with prescribed stress histories (Tables 4.1–4.2). However, it requires many inversions of the Jacobian matrix (as shown within parentheses in Tables 4.1 and 4.2). The size of the Jacobian matrix in these uniaxial problems is 3×3 but much larger matrices have to be inverted in multiaxial problems. Results from Gear's packages are not presented for the strain-controlled simulations (Tables 4.3 and 4.4) because in these cases these packages allowed large time-steps very early during integration, resulting in numerical instability and termination of the integration process.

The effect of error bounds e_{max} and e_{min} on the results of creep simulations is shown in Table 4.5. The 'optimal' (4 and 3) strategy has been used for these computations. The integration process is stable in all these cases and the creep strain after 80 hours varies little with error

TABLE 4.5
EFFECT OF ERROR BOUNDS ON THE ACCUMULATED NONELASTIC STRAIN IN CONSTANT STRESS CREEP OF 304 SS AT 500°C UP TO 80 HR

e_{max}	e_{min}	*Accumulated nonelastic strain*	*Number of function evaluations*
10^{-3}	5×10^{-4}	2·7015	932
10^{-3}	10^{-4}	2·7013	910
10^{-3}	5×10^{-5}	2·7042	925
10^{-3}	10^{-5}	2·7006	1 166
10^{-3}	5×10^{-6}	2·7052	1 321
5×10^{-4}	10^{-4}	2·7059	947
10^{-3}	10^{-4}	2·7013	910
5×10^{-3}	10^{-4}	2·7234	1 086
10^{-2}	10^{-4}	2·7005	1 113
10^{-4}	10^{-5}	2·7113	1 192
10^{-2}	10^{-3}	2·687	1 078

$\sigma = 42{\cdot}97$ ksi $\sigma_0^* = 38$ ksi

bounds. The values $e_{max} = 10^{-3}$ and $e_{min} = 10^{-4}$ seem optimal in the sense that these values minimize the number of function evaluations. Although the various strategies have been evaluated for this specific choice of error limits, it is expected that the broad conclusions regarding comparison of these methods will hold for other sensible choices of error limits.

The crucial question, of course, is the usefulness of a time-integration strategy for the solution of multiaxial boundary value problems. The predictor–corrector scheme and the one-step Euler method have been used to solve a variety of multiaxial problems in conjunction with the boundary element,[8–13] finite element[14,15] and finite difference methods.[16,17] The one-step Euler method has proved to be very simple and easy to use, yet very efficient for this class of problems.

REFERENCES

1. GEAR, C. W. The automatic integration of stiff ordinary differential equations. *Information Processing 68*, A. J. H. Morrell (ed.), North Holland, Amsterdam, 187–193 (1969).
2. GEAR, C. W. The automatic integration of ordinary differential equations, *Communications of the ACM*, **14**, 176–179 (1971).
3. GEAR, C. W. Algorithm 407 DIFSUB for solution of ordinary differential equations [D2], *Communications of the ACM*, **14**, 185–190 (1971).
4. CORMEAU, I. Numerical stability in quasi-static elastic/viscoplasticity. *International Journal for Numerical Methods in Engineering*, **9**, 109–127 (1975).
5. KUMAR, V., MORJARIA, M. and MUKHERJEE, S. Numerical integration of some stiff constitutive models of inelastic deformation, *American Society of Mechanical Engineers Journal of Engineering Materials and Technology*, **102**, 92–96 (1980).
6. HILDERBRAND, F. B. *Finite Difference Equations and Simulations*, Prentice Hall, New Jersey (1968).
7. KUMAR, V., MUKHERJEE, S., HUANG, F. H. and LI, C–Y. *Deformation in Type 304 Austenitic Stainless Steel*, Electric Power Research Institute Report EPRI NP-1276 for Project 697–1, Palo Alto, CA (1979).
8. MUKHERJEE, S. and KUMAR, V. Numerical analysis of time-dependent inelastic deformation in metallic media using the boundary-integral equation method. *American Society of Mechanical Engineers Journal of Applied Mechanics*, **45**, 785–790 (1978).
9. MORJARIA, M. and MUKHERJEE, S. Improved boundary-integral equation method for time-dependent inelastic deformation in metals. *International Journal for Numerical Methods in Engineering*, **15**, 97–111 (1980).
10. MORJARIA, M. and MUKHERJEE, S. Inelastic analysis of transverse deflection of plates by the boundary element method. *American Society of Mechanical Engineers Journal of Applied Mechanics*, **47**, 291–296 (1980).
11. MUKHERJEE, S. and MORJARIA, M. Comparison of boundary element and finite element methods in the inelastic torsion of prismatic shafts. *International Journal for Numerical Methods in Engineering*, **17**, 1576–1588 (1981).
12. MORJARIA, M. and MUKHERJEE, S. Numerical analysis of planar, time-dependent, inelastic deformation of plates with cracks by the boundary

element method. *International Journal of Solids and Structures*, **17**, 127–143 (1981).

13. MUKHERJEE, S. and MORJARIA, M. Boundary element analysis of time-dependent inelastic deformation of cracked plates loaded in anti-plane shear. *International Journal of Solids and Structures*, **17**, 753–763 (1981).
14. MORJARIA, M., SARIHAN, V. and MUKHERJEE, S. Comparison of boundary element and finite element methods in two-dimensional inelastic analysis. *Res Mechanica*, **1**, 3–20 (1980).
15. MORJARIA, M. and MUKHERJEE, S. Finite element analysis of time-dependent inelastic deformation in the presence of transient thermal stresses. *International Journal of Numerical Methods in Engineering*, **17**, 909–921 (1981).
16. MUKHERJEE, S., KUMAR, V. and CHANG, K. J. Elevated temperature inelastic analysis of metallic media under time varying loads using state variable theories. *International Journal of Solids and Structures*, **14**, 663–679 (1978).
17. MUKHERJEE, S. Thermoviscoplastic response of cylindrical structures using a state variable theory. *Mechanical Behavior of Materials—Proceedings of ICM 3*, Cambridge, England, K. J. Miller and R. F. Smith (eds.), Pergamon Press, Oxford and New York, **2**, 233–242 (1979).

CHAPTER 5

Planar Viscoplasticity Problems

Boundary element and finite element formulations for planar (plane strain and plane stress) viscoplasticity problems are presented in this chapter. Numerical results are presented for some illustrative plane stress problems and comparisons of boundary element, finite element and (whenever possible) direct solutions are carried out.

5.1 GOVERNING DIFFERENTIAL EQUATIONS

The equations for planar problems are obtained from the three-dimensional ones in the usual way. All variables are assumed to depend only on the in-plane coordinates x_1 and x_2 and time t. The nonzero components of stress are taken to be σ_{11}, σ_{22}, σ_{12} $(=\sigma_{21})$ and σ_{33} with the corresponding nonzero components of total strain being ε_{11}, ε_{22}, ε_{12} $(=\varepsilon_{21})$ and ε_{33}. Further, for plane strain, $u_3=\varepsilon_{33}=0$ and, for plane stress, $\sigma_{33}=0$. *The range of subscripts in all the equations in this chapter is 1, 2.*

The equilibrium kinematic and generalized Hooke's law retain the same forms as the three-dimensional ones (eqns. (3.1), (3.36) and (3.3) respectively). The Navier's equations for displacement rates, the starting point for the boundary element formulation, appear different for plane stress and plane strain. These are given below.

5.1.1 Plane Strain ($\varepsilon_{33}=0$)

The displacement rate equation in this case is

$$\dot{u}_{i,jj}+\frac{1}{1-2\nu}\dot{u}_{k,ki}=2\dot{\varepsilon}^{(n)}_{ij,j} \tag{5.1}$$

where physical body forces are absent and the incompressibility condition $\dot{\varepsilon}^{(n)}_{11}+\dot{\varepsilon}^{(n)}_{22}+\dot{\varepsilon}^{(n)}_{33}=0$ has been used.

The traction rate at a point on the surface ∂B of the body is

$$\dot{\tau}_i = G\left[(\dot{u}_{i,j} + \dot{u}_{j,i})n_j + \frac{2\nu}{1-2\nu}\dot{u}_{k,k}n_i - 2\dot{\varepsilon}_{ij}^{(n)}n_j \right] \tag{5.2}$$

These equations have the same form as those for three dimensions. The range of subscripts, of course, is reduced, as noted above.

5.1.2 Plane Stress ($\sigma_{33}=0$)

The displacement rate equation for this case, in the absence of physical body forces, is

$$\dot{u}_{i,jj} + \frac{1}{1-2\bar{\nu}}\dot{u}_{k,ki} = 2\dot{\varepsilon}_{ij,j}^{(n)} + \frac{2\bar{\nu}}{1-2\bar{\nu}}\dot{\varepsilon}_{kk,i}^{(n)} \tag{5.3}$$

where $\bar{\nu} = \nu/(1+\nu)$. It is interesting to note that replacement of ν by $\bar{\nu}$ is only valid for the elastic part and that $\dot{\varepsilon}_{kk}^{(n)} = \dot{\varepsilon}_{11}^{(n)} + \dot{\varepsilon}_{22}^{(n)} \neq 0$.

This time, the traction rate at a point on the surface ∂B is

$$\dot{\tau}_i = G\left\{ (\dot{u}_{i,j} + \dot{u}_{j,i})n_j + \frac{2\bar{\nu}}{1-2\bar{\nu}}\dot{u}_{k,k}n_i - 2\dot{\varepsilon}_{ij}^{(n)}n_j - \frac{2\bar{\nu}}{1-2\bar{\nu}}\dot{\varepsilon}_{kk}^{(n)}n_i \right\} \tag{5.4}$$

5.2 BOUNDARY ELEMENT FORMULATION FOR DISPLACEMENT RATES

5.2.1 Plane Strain ($\varepsilon_{33}=0$)

The integral equation for plane strain, for the displacement rate at an internal point p, can be written as (see eqn. (3.38))

$$\dot{u}_j(p) = \int_{\partial B} \{U_{ij}(p,Q)\dot{\tau}_i(Q) - T_{ij}(p,Q)\dot{u}_i(Q)\}\,\mathrm{d}c_Q + \int_B 2G U_{ij,k}(p,q)\dot{\varepsilon}_{ik}^{(n)}(q)\,\mathrm{d}A_q \tag{5.5}$$

where the kernels are given by Kelvin's singular solution for a point load in plane strain. These are

$$U_{ij} = -\frac{1}{8\pi(1-\nu)G}\{(3-4\nu)\ln r\,\delta_{ij} - r_{,i}r_{,j}\} \tag{5.6}$$

$$T_{ij} = -\frac{1}{4\pi(1-\nu)r}\left[\{(1-2\nu)\delta_{ij} + 2r_{,i}r_{,j}\}\frac{\partial r}{\partial n} + (1-2\nu)(r_{,i}n_j - r_{,j}n_i) \right] \tag{5.7}$$

Equation (5.5) is derived in a manner which is entirely analogous to the case of three dimensions.

In conformity with previous publications,[1,2] the area integral can be written as

$$2GU_{ij,k}\dot{\varepsilon}_{ik}^{(n)}=\frac{2G}{2}[U_{ij,k}+U_{kj,i}]\dot{\varepsilon}_{ik}^{(n)}=\Sigma_{ikj}\dot{\varepsilon}_{ik}^{(n)}$$

where

$$\Sigma_{ijk}=-\frac{1}{4\pi(1-v)r}\{(1-2v)(r_{,i}\delta_{jk}+r_{,j}\delta_{ki})$$
$$-r_{,k}\delta_{ij}+2r_{,i}r_{,j}r_{,k}\} \tag{5.8}$$

The boundary integral equation when the internal point p approaches a boundary point P, becomes[3,4]

$$c_{ij}(P)\dot{u}_i(P)=\int_{\partial B}\{U_{ij}(P,Q)\dot{\tau}_i(Q)-T_{ij}(P,Q)\dot{u}_i(Q)\}\mathrm{d}c_Q$$
$$+\int_B \Sigma_{ikj}(P,q)\dot{\varepsilon}_{ik}^{(n)}(q)\mathrm{d}A_q \tag{5.9}$$

where the components of the tensor c_{ij} are[3,4]

$$c_{11}=\frac{\beta}{2\pi}+\frac{\cos 2\gamma\sin\beta}{4\pi(1-v)}$$

$$c_{12}=c_{21}=-\frac{\sin 2\gamma\sin\beta}{4\pi(1-v)}$$

$$c_{22}=\frac{\beta}{2\pi}-\frac{\cos 2\gamma\sin\beta}{4\pi(1-v)} \tag{5.10}$$

In the above, $\dot{P}$ lies on a boundary corner of included angle β and γ is the angle between the bisector of β and the x_1 axis (Fig. 5.1). As usual, if $\beta=\pi$, $c_{11}=c_{22}=\frac{1}{2}$ and $c_{12}=0$ so that $c_{ij}=(\frac{1}{2})\delta_{ij}$.

5.2.2 Plane Stress ($\sigma_{33}=0$)

A pseudo body force F_i' and a pseudo boundary traction τ_i' are defined in

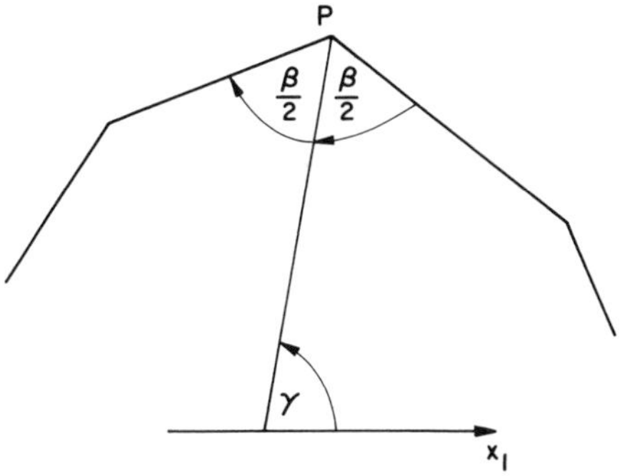

FIG. 5.1. Limiting procedure for an internal point approaching a boundary corner.

this case as

$$\dot{F}_i' = -2G\dot{\varepsilon}_{ij,j}^{(n)} - \frac{2G\bar{v}}{1-2\bar{v}}\dot{\varepsilon}_{kk,i}^{(n)} \tag{5.11}$$

$$\dot{\tau}_i' = \dot{\tau}_i + 2G\dot{\varepsilon}_{ij}^{(n)}n_j + \frac{2G\bar{v}}{1-2\bar{v}}\dot{\varepsilon}_{kk}^{(n)}n_i \tag{5.12}$$

Proceeding as in the case of three dimensions, this leads to the equation[5]

$$\dot{u}_j(p) = \int_{\partial B} \{U_{ij}(p,Q)\dot{\tau}_i'(Q) - T_{ij}(p,Q)\dot{u}_i(Q)\}\mathrm{d}c_Q + \int_B U_{ij}(p,q)\dot{F}_i'(q)\mathrm{d}A_q \tag{5.13}$$

where the elastic kernels U_{ij} and T_{ij} have the same form as eqns. (5.6) and (5.7) with v replaced by $\bar{v} = v/(1+v)$. A more convenient form of this equation can be obtained by substituting for F_i' and τ_i' from eqns (5.11) and (5.12) and using the divergence theorem to transform some area integrals into line integrals. This gives[1,2]

$$\dot{u}_j(p) = \int_{\partial B} \{U_{ij}(p,Q)\dot{\tau}_i(Q) - \mathrm{T}_{ij}(p,Q)\dot{u}_i(Q)\}\mathrm{d}c_Q + \int_B \Sigma_{ikj}(p,q)\dot{\varepsilon}_{ik}^{(n)}(q)\mathrm{d}A_q \tag{5.14}$$

where

$$\Sigma_{ijk}=G\left\{U_{ik,j}+U_{kj,i}+\frac{2\bar{v}}{1-2\bar{v}}U_{lk,l}\delta_{ij}\right\}$$

$$=-\frac{1}{4\pi(1-\bar{v})r}\{(1-2\bar{v})(r_{,i}\delta_{jk}+r_{,j}\delta_{ki}-r_{,k}\delta_{ij})$$

$$+2r_{,i}r_{,j}r_{,k}\} \tag{5.15}$$

The kernel Σ_{ijk} is related to the stress σ^*_{ij} due to a point force in a two-dimensional elastic body in plane stress by the equation (see eqns (3.7)–(3.9))

$$\sigma^*_{ij}=\Sigma_{ijk}e_k \tag{5.16}$$

The boundary integral equation when p approaches P on ∂B is

$$c_{ij}(P)\dot{u}_i(P)=\int_{\partial B}\{U_{ij}(P,Q)\dot{\tau}_i(Q)-T_{ij}(P,Q)\dot{u}_i(Q)\}\mathrm{d}c_Q$$

$$+\int_B\Sigma_{ijk}(P,q)\dot{\varepsilon}^{(n)}_{ik}(q)\mathrm{d}A_q \tag{5.17}$$

The components of the tensor c_{ij} are of the same form as eqn. (5.10) with v replaced by $\bar{v}$. This is so because the tensor c_{ij} is the residue, obtained by the limiting process, from the kernel T_{ij}.

5.3 STRESS RATES

The stress rates can be determined from the displacement rates by using any one of the techniques described earlier for three-dimensional problems in Section 3.3.3.

5.3.1 Plane Strain

The displacement rate gradient can be obtained analytically by differentiating eqn. (5.5) at a source point, to give

$$\dot{u}_{j,L}(p)=\int_{\partial B}\{U_{ij,L}(p,Q)\dot{\tau}_i(Q)-T_{ij,L}(p,Q)\dot{u}_i(Q)\}\mathrm{d}c_Q$$

$$+\frac{\partial}{\partial x_L}\int_B\Sigma_{ikj}(p,q)\dot{\varepsilon}^{(n)}_{ik}(q)\mathrm{d}A_q \tag{5.18}$$

where, for reasons mentioned earlier, the kernel Σ_{ikj} cannot be differentiated directly under the integral sign. Now the stress rate is given by

$$\dot{\sigma}_{ij}=G\{(\dot{u}_{i,j}+\dot{u}_{j,i})+\frac{2\nu}{1-2\nu}\dot{u}_{k,k}\delta_{ij}-2\dot{\varepsilon}_{ij}^{(n)}\} \tag{5.19}$$

5.3.2 Plane Stress

The displacement rate gradient can be obtained from an expression identical to the right-hand side of eqn (5.18) where the appropriate kernels for plane stress are used. In the plane stress numerical examples presented later in this chapter, the area integral in eqn. (5.18) is evaluated analytically for an arbitrary source point p and is then differentiated at p.[6,4] This will be discussed in more detail in Section 5.8. Finally, the stress rate equation in terms of displacement rates is

$$\dot{\sigma}_{ij}=G\left\{(\dot{u}_{i,j}+\dot{u}_{j,i})+\frac{2\bar{\nu}}{1-2\bar{\nu}}\dot{u}_{k,k}\delta_{ij}-2\dot{\varepsilon}_{ij}^{(n)}-\frac{2\bar{\nu}}{1-2\bar{\nu}}\dot{\varepsilon}_{kk}^{(n)}\delta_{ij}\right\} \tag{5.20}$$

5.4 AN ANALYTICAL EXAMPLE FOR PLANE STRAIN

The analytical example considered here is that of a long thick-walled cylinder of internal and external radii a and b subjected to constant internal and external pressures p_i and p_o, respectively. The cylinder is assumed to be long and plane strain conditions ($\varepsilon_{33}=0$) are assumed to prevail sufficiently far from the ends. Plane cross-sections are assumed to remain plane. The radial displacement component u is a function of R and t only and $u_\theta=0$. The nonvanishing stress and strain components, σ_{RR}, $\sigma_{\theta\theta}$, σ_{zz}, ε_{RR} and $\varepsilon_{\theta\theta}$ are also functions of R and t only. Polar coordinates R, θ and z are used in this analysis.

A unit load is applied in the x_1 direction at a point on the x_1 axis as shown in Fig. 5.2. The integral eqns. (5.5) (with Σ_{ikj}) and (5.9) for the displacement rate assume the form (ρ or θ not summed)

$$\kappa\dot{u}(R)=a\dot{u}(a)\int_0^{2\pi}T_{\rho 1}(R;a,\theta)\mathrm{d}\theta-b\dot{u}(b)\int_0^{2\pi}T_{\rho 1}(R;b,\theta)\mathrm{d}\theta$$
$$+\int_a^b\int_0^{2\pi}\{\Sigma_{\rho\rho 1}(R;\rho,\theta)\dot{\varepsilon}_{\rho\rho}^{(n)}(\rho)$$
$$+\Sigma_{\theta\theta 1}(R;\rho,\theta)\dot{\varepsilon}_{\theta\theta}^{(n)}(\rho)\}\rho\mathrm{d}\theta\mathrm{d}\rho \tag{5.21}$$

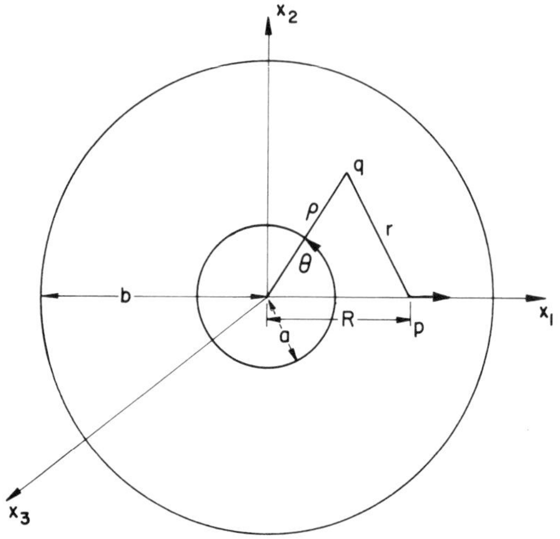

FIG. 5.2. Geometry of the cylinder and disc problems.

where $\kappa=1$ for an interior point and $\kappa=\frac{1}{2}$ for a boundary point. The kernels, in terms of the distance r between a source point and a field point, are

$$U_{\rho 1}(R;\rho,\theta)=-\frac{1}{8\pi(1-v)G}\left\{(3-4v)(\ln r)\cos\theta - \frac{(\rho\cos\theta-R)(\rho-R\cos\theta)}{r^2}\right\} \tag{5.22}$$

$$T_{\rho 1}(R;\rho,\theta)=-\frac{1}{4\pi(1-v)}\left[(1-2v)\left\{\frac{2(\rho-R\cos\theta)\cos\theta}{r^2}-\frac{(\rho\cos\theta-R)}{r^2}\right\}+\frac{2(\rho\cos\theta-R)(\rho-R\cos\theta)^2}{r^4}\right] \tag{5.23}$$

$$\Sigma_{\rho\rho 1}(R;\rho,\theta)=-\frac{1}{4\pi(1-v)}\left[(1-2v)\left\{\frac{2(\rho-R\cos\theta)\cos\theta}{r^2}\right\}-\frac{(\rho\cos\theta-R)}{r^2}+\frac{2(\rho\cos\theta-R)(\rho-R\cos\theta)^2}{r^4}\right]$$

$$\Sigma_{\theta\theta 1}(R;\rho,\theta) = -\frac{1}{4\pi(1-\nu)}\left\{-(1-2\nu)\frac{2R\sin^2\theta}{r^2} - \frac{(\rho\cos\theta - R)}{r^2}\right.$$

$$\left. + \frac{2(\rho\cos\theta - R)R^2\sin^2\theta}{r^4}\right\} \tag{5.24}$$

$$r^2 = \rho^2 + R^2 - 2\rho R\cos\theta$$

Using $\dot{\varepsilon}^{(n)}_{RR} + \dot{\varepsilon}^{(n)}_{\theta\theta} + \dot{\varepsilon}^{(n)}_{zz} = 0$, the double integral in eqn. (5.21) reduces to

$$-\frac{1-2\nu}{2(1-\nu)}\left\{\frac{1}{R}\int_a^R \rho\dot{\varepsilon}^{(n)}_{zz}\,\mathrm{d}\rho + R\int_r^b \frac{(\dot{\varepsilon}^{(n)}_{RR} - \dot{\varepsilon}^{(n)}_{\theta\theta})}{\rho}\,\mathrm{d}\rho\right\}.$$

The appropriate integrals are listed in Table 5.1.

TABLE 5.1
INTEGRALS FOR CYLINDER AND DISC PROBLEMS (REFERENCE 16 HAS BEEN USED TO EVALUATE THESE INTEGRALS)

$f(\theta)$	$\int_0^{2\pi} f(\theta)\mathrm{d}\theta$		
	$\rho > R$	$\rho = R$	$\rho < R$
$\cos\theta\ln r$	$-\frac{\pi R}{\rho}$	$-\pi$	$-\frac{\pi\rho}{R}$
$\frac{(\rho - R\cos\theta)(\rho\cos\theta - R)}{r^2}$	$-\frac{\pi R}{\rho}$	$-\pi$	$-\frac{\pi\rho}{R}$
$\frac{2\cos\theta(\rho - R\cos\theta)}{r^2}$	$\frac{2\pi R}{\rho^2}$	0	$-\frac{2\pi}{R}$
$\frac{2(\rho\cos\theta - R)(\rho - R\cos\theta)^2}{r^4}$	0	$-\frac{\pi}{R}$	$-\frac{2\pi}{R}$
$\frac{\rho\cos\theta - R}{r^2}$	0	$-\frac{\pi}{R}$	$-\frac{2\pi}{R}$
$\frac{2R\sin^2\theta}{r^2}$	$\frac{2\pi R}{\rho^2}$	$\frac{2\pi}{R}$	$\frac{2\pi}{R}$
$\frac{2(\rho\cos\theta - R)R^2\sin^2\theta}{r^4}$	0	$-\frac{\pi}{R}$	$-\frac{2\pi}{R}$

$$r^2 = \rho^2 + R^2 - 2\rho R\cos\theta$$

In order to solve for the displacement, eqn. (5.21) is used with $R=a$ and then with $R=b$. This gives two linear algebraic equations for the boundary displacement rates $\dot{u}(a)$ and $\dot{u}(b)$. Solving these,

$$\dot{u}(a)=-\frac{ab^2}{b^2-a^2}\int_a^b\frac{(\dot{\varepsilon}_{RR}^{(n)}-\dot{\varepsilon}_{\theta\theta}^{(n)})}{\rho}\mathrm{d}\rho - \frac{a(1-2v)}{b^2-a^2}\int_a^b\rho\dot{\varepsilon}_{zz}^{(n)}\mathrm{d}\rho \tag{5.25}$$

$$\dot{u}(b)=-\frac{a^2b}{b^2-a^2}\int_a^b\frac{(\dot{\varepsilon}_{RR}^{(n)}-\dot{\varepsilon}_{\theta\theta}^{(n)})}{\rho}\mathrm{d}\rho - \frac{b(1-2v)}{b^2-a^2}\int_a^b\rho\dot{\varepsilon}_{zz}^{(n)}\mathrm{d}\rho. \tag{5.26}$$

Next, eqn. (5.21) is used for an interior point to give

$$\dot{u}(R)=\frac{1}{2(1-v)}\left[(1-2v)R\int_a^R\frac{(\dot{\varepsilon}_{RR}^{(n)}-\dot{\varepsilon}_{\theta\theta}^{(n)})}{\rho}\mathrm{d}\rho - \frac{b^2}{b^2-a^2}\left\{(1-2v)R+\frac{a^2}{R}\right\}\int_a^b\frac{(\dot{\varepsilon}_{RR}^{(n)}-\dot{\varepsilon}_{\theta\theta}^{(n)})}{\rho}\mathrm{d}\rho\right] - \frac{1-2v}{2(1-v)}\left[\frac{1}{R}\int_a^R\rho\dot{\varepsilon}_{zz}^{(n)}\mathrm{d}\rho+\frac{1}{b^2-a^2}\left\{(1-2v)R+\frac{a^2}{R}\right\}\int_a^b\rho\dot{\varepsilon}_{zz}^{(n)}\mathrm{d}\rho\right]. \tag{5.27}$$

Finally the stress rates are obtained using Hooke's law written in cylindrical coordinates. These are

$$\dot{\sigma}_{RR}=\frac{E}{2(1-v^2)}\left\{\int_a^R\frac{(\dot{\varepsilon}_{RR}^{(n)}-\dot{\varepsilon}_{\theta\theta}^{(n)})}{\rho}\mathrm{d}\rho-\frac{(R-a^2)b^2}{(b^2-a^2)R^2}\int_a^b\frac{(\dot{\varepsilon}_{RR}^{(n)}-\dot{\varepsilon}_{\theta\theta}^{(n)})}{\rho}\mathrm{d}\rho\right\} + \frac{E(1-2v)}{2(1-v^2)}\frac{1}{R^2}\left\{\int_a^R\rho\dot{\varepsilon}_{zz}^{(n)}\mathrm{d}\rho-\frac{(R^2-a^2)}{(b^2-a^2)}\int_a^b\rho\dot{\varepsilon}_{zz}^{(n)}\mathrm{d}\rho\right\} \tag{5.28}$$

$$\dot{\sigma}_{\theta\theta}=\frac{E}{2(1-v^2)}\left\{\int_a^R\frac{(\dot{\varepsilon}_{RR}^{(n)}-\dot{\varepsilon}_{\theta\theta}^{(n)})}{\rho}\mathrm{d}\rho-\frac{(R^2+a^2)b^2}{(b^2-a^2)R^2}\int_a^b\frac{(\dot{\varepsilon}_{RR}^{(n)}-\dot{\varepsilon}_{\theta\theta}^{(n)})}{\rho}\mathrm{d}\rho\right\} - \frac{E(1-2v)}{2(1-v^2)}\frac{1}{R^2}\left\{\int_a^R\rho\dot{\varepsilon}_{zz}^{(n)}\mathrm{d}\rho+\frac{(R^2+a^2)}{(b^2-a^2)}\int_a^b\rho\dot{\varepsilon}_{zz}^{(n)}\mathrm{d}\rho\right\} - \frac{E}{(1-v^2)}(\dot{\varepsilon}_{\theta\theta}^{(n)}+v\dot{\varepsilon}_{zz}^{(n)}) \tag{5.29}$$

$$\dot{\sigma}_{zz}=\frac{E\nu}{1-\nu^2}\left\{\int_a^R\frac{(\dot{\varepsilon}_{RR}^{(n)}-\dot{\varepsilon}_{\theta\theta}^{(n)})}{\rho}\mathrm{d}\rho-\frac{b^2}{(b^2-a^2)}\int_a^b\frac{(\dot{\varepsilon}_{RR}^{(n)}-\dot{\varepsilon}_{\theta\theta}^{(n)})}{\rho}\mathrm{d}\rho\right\}$$

$$-\frac{E\nu(1-2\nu)}{(1-\nu^2)}\frac{1}{(b^2-a^2)}\int_a^b\rho\dot{\varepsilon}_{zz}^{(n)}\mathrm{d}\rho+\frac{E}{1-\nu^2}\left\{\nu\dot{\varepsilon}_{RR}^{(n)}-(1-\nu)\dot{\varepsilon}_{zz}^{(n)}\right\}. \tag{5.30}$$

These formulae were obtained in an entirely different manner (directly from the governing differential equations in cylindrical coordinates) in reference 7.

5.5 AN ANALYTICAL EXAMPLE FOR PLANE STRESS

This example concerns a thin uniform circular disc of internal and external radii a and b, thickness h, rotating with a constant angular velocity ω about its axis of symmetry (Fig. 5.2). The surfaces $R=a$ and $R=b$ are traction free. Plane stress conditions are assumed to prevail so that $\sigma_{zz}=0$. Under these assumptions, and because of axisymmetry, $u_\theta=0$ and the radial displacement is a function of R and t only. The nonvanishing stresses and strains are σ_{RR}, $\sigma_{\theta\theta}$, ε_{RR}, $\varepsilon_{\theta\theta}$ and ε_{zz} and these are functions of R and t only. The mass density of the disc material is m.

The procedure is analogous to that of the plane strain case. A unit load is applied in the x_1 direction at the point $(R, 0, 0)$ on the x_1 axis as shown in Fig. 5.2. The displacement rate is governed by eqn. (5.21). The kernels $U_{\rho 1}$ and $T_{\rho 1}$ have the same form as eqns. (5.22) and (5.23) with ν replaced by $\bar{\nu}=\nu/(1+\nu)$. The kernels for the double integral are

$$\Sigma_{\rho\rho 1}(R;\rho,\theta)=-\frac{1}{4\pi(1-\bar{\nu})}\left[(1-2\bar{\nu})\left\{\frac{2(\rho-R\cos\theta)\cos\theta}{r^2}-\frac{(\rho\cos\theta-R)}{r^2}\right\}+\frac{2(\rho\cos\theta-R)(\rho-R\cos\theta)^2}{r^4}\right] \tag{5.31}$$

$$\Sigma_{\theta\theta 1}(R;\rho,\theta)=-\frac{1}{4\pi(1-\bar{\nu})}\left[-(1-2\bar{\nu})\left\{\frac{2R\sin^2\theta}{r^2}+\frac{(\rho\cos\theta-R)}{r^2}\right\}+\frac{2(\rho\cos\theta-R)R^2\sin^2\theta}{r^4}\right] \tag{5.32}$$

with $\quad r^2=\rho^2+R^2-2\rho R\cos\theta.$

This time the double integral in eqn. (5.21) becomes

$$-\frac{1}{2(1-\bar{\nu})}\left\{\frac{1}{R}\int_a^R \rho\dot{\varepsilon}_{zz}^{(n)}\mathrm{d}\rho+(1-2\bar{\nu})R\int_r^b \frac{(\dot{\varepsilon}_{RR}^{(n)}-\dot{\varepsilon}_{\theta\theta}^{(n)})}{\rho}\mathrm{d}\rho\right\}. \quad (5.33)$$

Once again $\dot{u}(a)$ and $\dot{u}(b)$ and then $\dot{u}(R)$ are determined as before. The displacement rate is

$$\dot{u}(R)=\frac{1}{2(1-\bar{\nu})}\left[(1-2\bar{\nu})R\int_a^R \frac{(\dot{\varepsilon}_{RR}^{(n)}-\dot{\varepsilon}_{\theta\theta}^{(n)})}{\rho}\mathrm{d}\rho-\frac{b^2}{b^2-a^2}\right.$$

$$\times\left\{(1-2\bar{\nu})R+\frac{a^2}{R}\right\}\int_a^b \frac{(\dot{\varepsilon}_{RR}^{(n)}-\dot{\varepsilon}_{\theta\theta}^{(n)})}{\rho}\mathrm{d}\rho$$

$$\left.-\frac{1}{R}\int_a^R \rho\dot{\varepsilon}_{zz}^{(n)}\mathrm{d}\rho-\frac{1}{b^2-a^2}\left\{(1-2\bar{\nu})R+\frac{a^2}{R}\right\}\int_a^b \rho\dot{\varepsilon}_{zz}^{(n)}\mathrm{d}\rho\right]. \quad (5.34)$$

The equations for the stress rates $\dot{\sigma}_{RR}$ and $\dot{\sigma}_{\theta\theta}$ have exactly the same form as eqns. (5.28) and (5.29) for plane strain with *the Poisson's ratio* ν *set equal to zero*. The axial strain rate is now

$$\dot{\varepsilon}_{zz}=-\nu\left\{\int_a^R \frac{(\dot{\varepsilon}_{RR}^{(n)}-\dot{\varepsilon}_{\theta\theta}^{(n)})}{\rho}\mathrm{d}\rho-\frac{b^2}{(b^2-a^2)}\int_a^b \frac{\dot{\varepsilon}_{RR}^{(n)}-\dot{\varepsilon}_{\theta\theta}^{(n)}}{\rho}\mathrm{d}\rho\right\}$$

$$+\frac{\nu}{(b^2-a^2)}\int_a^b \rho\dot{\varepsilon}_{zz}^{(n)}\mathrm{d}\rho+\dot{\varepsilon}_{zz}^{(n)}+\nu\dot{\varepsilon}_{\theta\theta}^{(n)} \quad (5.35)$$

These formulae were derived directly from the governing differential equations in cylindrical coordinates in reference 7.

5.6 FINITE ELEMENT FORMULATION

A simple finite element formulation for the problem can be based on a rate form of the principle of virtual work which can be written as

$$\int_B \{\delta\dot{\varepsilon}\}^T\{\dot{\sigma}\}\mathrm{d}A=\int_{\partial B}\{\delta\dot{u}\}^T\{\dot{\tau}\}\mathrm{d}c \quad (5.36)$$

where $\{\dot{\sigma}\}$ and $\{\dot{\tau}\}$ are the rates of stress and prescribed surface traction and $\{\delta\varepsilon\}$ and $\{\delta u\}$ are virtual strain and displacement vectors, respectively.

This leads to the equilibrium equation[8,9]

$$[K]\{\dot{\delta}\}+\{\dot{F}^{(n)}\}+\{\dot{F}^{(B)}\}=0 \tag{5.37}$$

where
$$[K]=\sum\int_{E}[B]^{T}[D][B]\mathrm{d}A$$

$$\{\dot{F}^{(n)}\}=-\sum\int_{E}[B]^{T}[D]\{\dot{\varepsilon}^{(n)}\}\mathrm{d}A$$

and
$$\{\dot{F}^{(B)}\}=-\sum\int_{\partial E}[N]^{T}\{\dot{\tau}\}\mathrm{d}c$$

In the above equations, $[K]$ is the stiffness matrix, $\{\dot{F}^{(n)}\}$ is the pseudonodal force due to the nonelastic strain rates and $\{\dot{F}^{(B)}\}$ is the equivalent nodal force due to the prescribed traction rate $\dot{\tau}$ on the boundary of the element. The area integrals are evaluated over the region of the element and the line integral is evaluated over that part of the boundary of the element on which the traction is prescribed. The vector $\{\dot{\delta}\}$ denotes nodal displacement rates which are related to the displacement rate field through the shape function matrix $[N]$.

$$\{\dot{u}\}=[N]\{\dot{\delta}\} \tag{5.38}$$

The local strain displacement relation in terms of nodal displacement has the form

$$\{\dot{\varepsilon}\}=[B]\{\dot{\delta}\} \tag{5.39}$$

and Hooke's law relating the elastic strain rates to the stress rates becomes

$$\{\dot{\sigma}\}=[D]\{\dot{\varepsilon}^{(e)}\}=[D](\{\dot{\varepsilon}\}-\{\dot{\varepsilon}^{(n)}\}) \tag{5.40}$$

5.7 NUMERICAL IMPLEMENTATION OF BEM FOR PLANE STRESS

The general strategy for the numerical implementation of the plane stress equations is quite analogous to that discussed earlier for three-dimensional problems in Section 3.3.6. The first step is the discretization of the two-dimensional body into boundary elements and internal cells. A discretized version of the boundary integral eqn. (5.17) for the displacement rate is

$$c_{ij}(P_M)\dot{u}_i(P_M)=\sum_{N_s}\int_{\Delta c_N} U_{ij}(P_M,Q)\dot{\tau}_i(Q)\mathrm{d}c_Q$$
$$-\sum_{N_s}\int_{\Delta c_N} T_{ij}(P_M,Q)\dot{u}_i(Q)\mathrm{d}c_Q$$
$$+\sum_{n_i}\int_{\Delta A_n} \Sigma_{ikj}(P_M,q)\dot{\varepsilon}_{ik}^{(n)}(q)\mathrm{d}Aq \qquad (5.41)$$

where the boundary of the body ∂B is divided into N_s boundary segments and the interior into n_i internal cells and $\dot{u}_i(P_M)$ are the components of the displacement rates at a point P which coincides with node M.

Suitable shape functions must now be chosen for the variation of traction and displacement rates on the surface elements Δc_N and the variation of the nonelastic strain rates over an internal cell ΔA_n. As mentioned before, integrals of kernels over elements on which they become singular must be obtained carefully. For planar problems with straight boundary elements and internal cells with straight boundaries, with fairly simple shape functions, it is possible to evaluate integrals of kernels analytically. This has been done in order to obtain the numerical results for planar viscoplastic problems published to date, and is recommended whenever possible.

A suitable strategy must be used for numerical modelling of possible jumps in normals or prescribed tractions across boundaries of boundary elements. A 'zero length' element placed at a corner is convenient for this purpose.

Numerical discretization transforms eqn. (5.41) into an algebraic system of the type

$$[A]\{\dot{u}\}+[B]\{\dot{\tau}\}=\{\dot{b}\} \qquad (5.42)$$

where the coefficient matrices $[A]$ and $[B]$ contain integrals of the kernels and the shape functions and the vector $\{\dot{b}\}$ contains the nonelastic strain rates and the kernel Σ_{ikj}. This equation, with $\{\dot{b}\}$ known at any time through the constitutive equations, is used to solve for the unspecified components of traction and displacement rates in terms of the specified ones. Once all the displacement and traction rates at any time are determined over the entire boundary, a discretized version of eqn. (5.1.4) is used to calculate displacement rates over internal points in the body.

The stress rates must now be determined throughout the body from the displacement rates. One strategy is the use of the plane stress version of eqn. (3.42) and then Hooke's law to determine stress rates. Another is to differentiate the kernels U_{ij} and T_{ij} in eqn. (5.14) under the integral sign but evaluate the area integral analytically over an internal cell for a source point p_m and then differentiate this term at p_m. Both these strategies have been used by researchers and some details of these are presented in the next section.

5.8 NUMERICAL RESULTS FOR PLANE STRESS PROBLEMS —COMPARISON OF BEM AND FEM RESULTS

The author and his group at Cornell University have published several papers in recent years containing numerical results for planar viscoplasticity problems obtained by using the boundary element method.[6,4,8,10] Other recent related work that has come to the author's attention is that due to Chaudonneret[11] and Telles and Brebbia.[12] The research reported in these papers concerns numerical results for plane stress problems. Various planar geometries and loading histories are considered. The constitutive models used to determine nonelastic strains are different in these papers. Mukherjee and his group use Hart's model in references 6, 4, 8, 10, Chaudonneret uses a model due to Lemaitre[13] and Telles and Brebbia one due to Perzyna.[14] An initial strain formulation is used by Mukherjee's group while an initial stress formulation is used by the others. The kernels are integrated analytically in all cases. Mukherjee's group uses straight boundary elements and polygonal internal cells with either a piecewise uniform[6] or piecewise linear[4,8] approximation of tractions and displacements on the boundary elements and piecewise uniform nonelastic strain rates on the internal cells. Chaudonneret also uses piecewise linear approximations of boundary variables on straight boundary elements and piecewise uniform nonelastic strain rates on rectangular internal cells. Telles and Brebbia, on the other hand, use piecewise linear nonelastic strain rates on internal cells. The strategy used for the determination of stress rates is pointwise (analytical differentiation of displacement rates) in all cases—an equation analogous to (3.42) by Chaudonneret and Telles and Brebbia and integration over an internal cell and then differentiation (as described below) by Mukherjee and his coworkers. Different time integration strategies are used by the different groups. The Euler formula with automatic time-step control, described in Chapter 4, is used by Mukherjee *et al.* in references 4 and 8. All the authors carry out comparisons of BEM and FEM results.

The numerical results obtained by the author's group are described in some detail in the rest of the chapter. These results (from references 4 and 8) are obtained by using piecewise linear approximations for boundary variables on straight boundary elements and uniform nonelastic strain rates on polygonal internal cells. The values of boundary variables are assigned at nodes which lie at the intersection of boundary segments. Possible discontinuities in tractions are taken care of by placing a 'zero length' element between two nodes and assigning different values of traction at each of these nodes. Nodes are allowed to lie on corners where the residue terms from eqns. (5.10) (with v replaced by $\bar{v}$) are used.

The explicit form of the discretized stress rate equation is given next. This is obtained from eqns. (5.14) and (5.20) as

$$
\begin{aligned}
\dot{\sigma}_{ij}(p_m) = & \sum_{N_s} \int_{\Delta c_N} V_{ijk}(p_M, Q)\dot{\tau}_k(Q)\,\mathrm{d}c_Q \\
& - \sum_{N_s} \int_{\Delta c_N} T_{ijk}(p_M, Q)\dot{u}_k(Q)\,\mathrm{d}c_Q \\
& + \sum_{n_i} \dot{\varepsilon}_{kl}^{(n)}(q_n)\Delta\Sigma_{ijkl}(p_m, q_n) \\
& - 2G\dot{\varepsilon}_{ij}^{(n)}(p_m) - \frac{2G\bar{v}}{1-2\bar{v}}\dot{\varepsilon}_{kk}^{(n)}(p_m)\delta_{ij}
\end{aligned}
\tag{5.43}
$$

where

$$
\begin{aligned}
\Delta\Sigma_{ijkl}(p_m, q_n) = G\Bigg\{ & \frac{\partial \Delta\Sigma_{kli}(p_m, q_n)}{\partial x_j|p_m} + \frac{\partial \Delta\Sigma_{klj}(p_m, q_n)}{\partial x_i|p_m} \\
& + \frac{2\bar{v}}{1-2\bar{v}} \frac{\partial \Delta\Sigma_{klr}(p_m, q_n)}{\partial x_r|p_m}\delta_{ij}\Bigg\}
\end{aligned}
$$

$$
\Delta\Sigma_{ijk}(p_m, q_n) = \int_{\Delta A_n} \Sigma_{ijk}(p_m, q)\,\mathrm{d}A_q
$$

$$
V_{ijk} = -\Sigma_{ijk}
$$

and

$$
T_{ijk} = -G\left\{ T_{ki,j} + T_{kj,i} + \frac{2\bar{v}}{1-2\bar{v}} T_{kl,l}\delta_{ij} \right\}
$$

$$
=\frac{G}{2\pi(1-\bar{\nu})r^2}\left[\{2(1-2\bar{\nu})\delta_{ij}r_{,k}+2\bar{\nu}(\delta_{jk}r_{,i}+\delta_{ki}r_{,j})-8r_{,i}r_{,j}r_{,k}\}\frac{\partial r}{\partial n}\right.
$$

$$
+2(1-2\bar{\nu})r_{,i}r_{,j}n_k+2\bar{\nu}(r_{,j}r_{,k}n_i+r_{,k}r_{,i}n_j)
$$

$$
\left.+(1-2\bar{\nu})(\delta_{ik}n_j+\delta_{jk}n_i)-(1-4\bar{\nu})\delta_{ij}n_k\right]
$$

It is important to note that while U_{ij} and T_{ij} can be differentiated under the integral sign, Σ_{ijk} is first integrated over an internal cell for an arbitrary source point p_m and is then differentiated at p_m. Details of the integration procedure are given in Section 5.9.

The finite element solutions reported below are obtained by using linear strain triangles (LST). The area integrals for $\{\dot{F}^{(n)}\}$ (eqn. (5.37)) are evaluated by Gaussian quadrature (7 Gauss points).

5.8.1 Parameter Values

The values of the parameters used in the numerical calculations, at 400 °C, are those given in Section 4.3. The values at 200 °C are available in reference 15. The initial values of the state variables for an annealed sample are

$$
\sigma^*(\mathbf{x},0)=17\text{ ksi}
$$

$$
\varepsilon_{ij}^{(a)}(\mathbf{x},0)=0
$$

5.8.2 Numerical Results

The calculated uniaxial creep response of a plate subjected to a constant applied stress of 20 ksi is shown in Fig. 5.3. The direct solution is obtained by numerical integration of the one-dimensional equations.[15] The results from the three methods coincide. Results for uniaxial constant strain rate extension at a strain rate of 0·02/min are shown in Fig. 5.4. Again the results from the three methods very nearly coincide. Only a quarter of the plate is modelled here because of symmetry. Eight boundary nodes and one internal cell are used in the BEM and two finite elements are used for the FEM calculations as shown.

Figures 5.5 and 5.6 show the BEM and FEM mesh, respectively, used to model a quarter of a circular plate with a concentric circular cutout under internal pressure increasing at a constant rate. The BEM calculation uses 30 boundary nodes and 20 internal cells while the FEM calculation has 48 elements and 117 nodes. Comparison of results is shown in Fig. 5.7. The BEM result is very close to the direct calculation

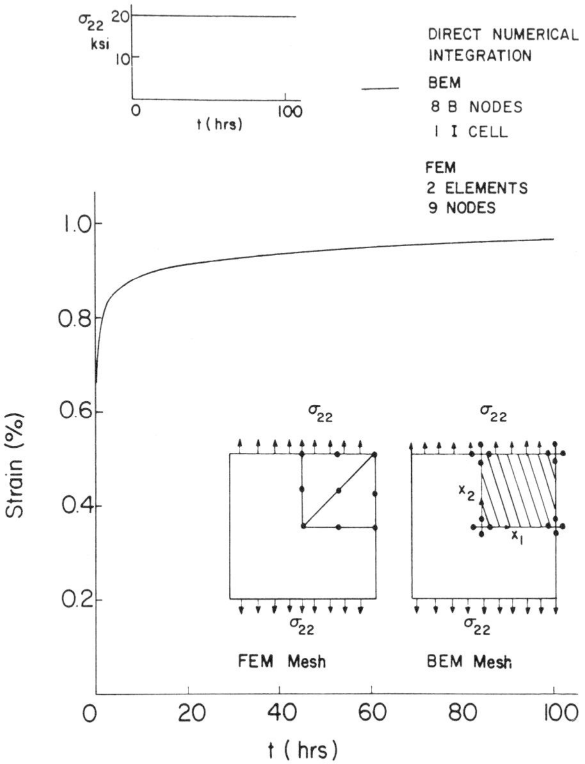

FIG. 5.3. Comparison of direct, BEM and FEM solutions for uniaxial creep at a constant stress of 20 ksi in an annealed 304 SS plate at 400° C.

while the FEM simulation has a maximum error of about 5% compared to the direct solution. For both the BEM and FEM calculations, the plane stress, as opposed to the axisymmetric option, has been used. The FEM mesh has fewer elements along the curved surface of the plate (compared to the BEM mesh) but has a linear (as opposed to uniform) description of strain on each internal element. The direct solution for this problem is obtained by the method discussed in reference 7.

The last problem considered is the inelastic response of a square plate with a circular cutout subjected to a constant remote uniaxial tensile stress of 10 ksi. Again, a quarter of the plate is modelled to take advantage of symmetry. The BEM mesh, shown in Fig. 5.8, has 37 boundary nodes and 20 internal cells while the FEM mesh, shown in Fig. 5.9, has 102 nodes and 41 elements.

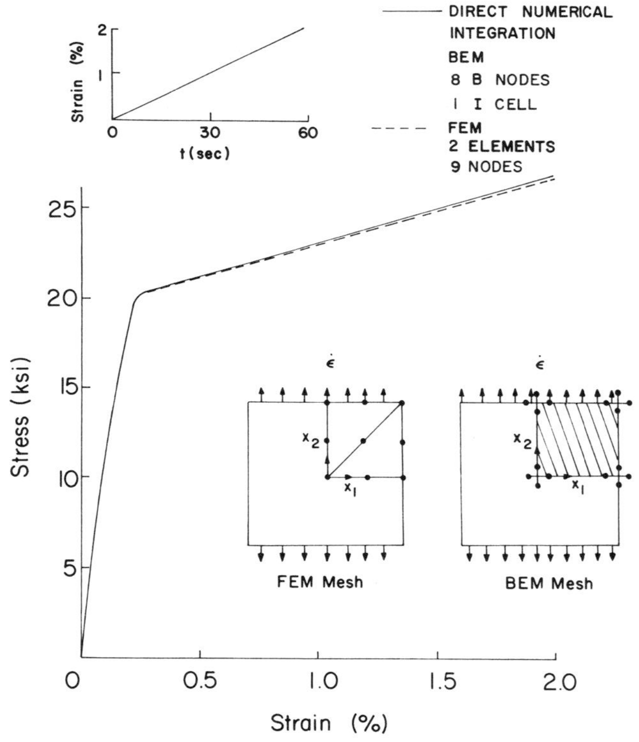

FIG. 5.4. Comparison of direct, BEM and FEM solutions for constant strain-rate extension in an annealed 304 SS plate at 400°C, $\dot{\varepsilon}=0{\cdot}02/\text{min}$.

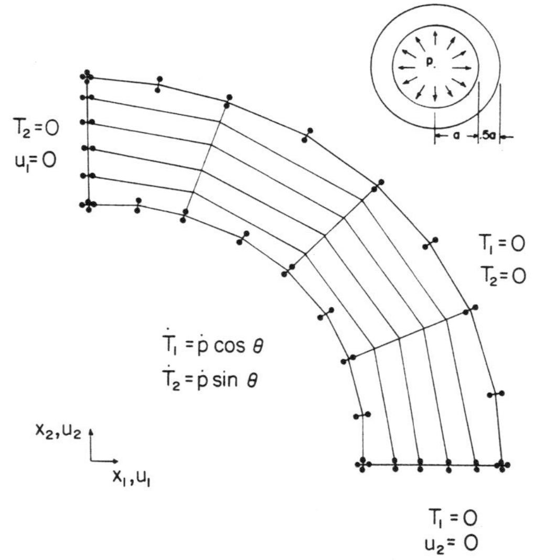

FIG. 5.5. BEM mesh for circular plate with concentric circular cutout under internal pressure. 30 boundary nodes and 20 internal cells.

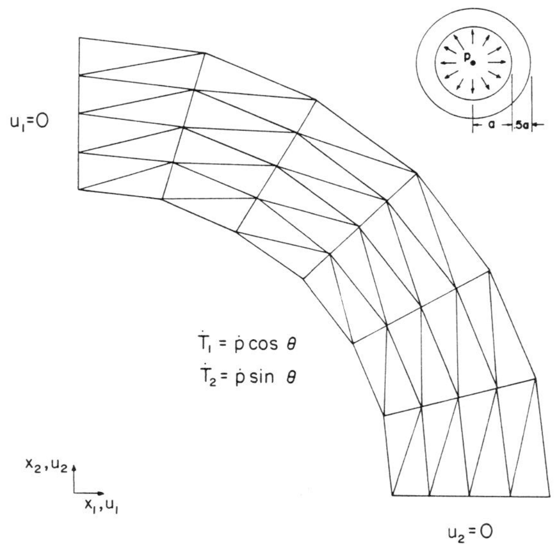

FIG. 5.6. FEM mesh for circular plate with concentric circular cutout under internal pressure. 117 nodes, 48 elements.

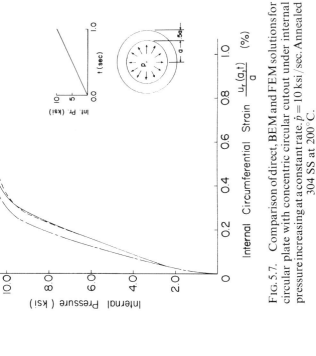

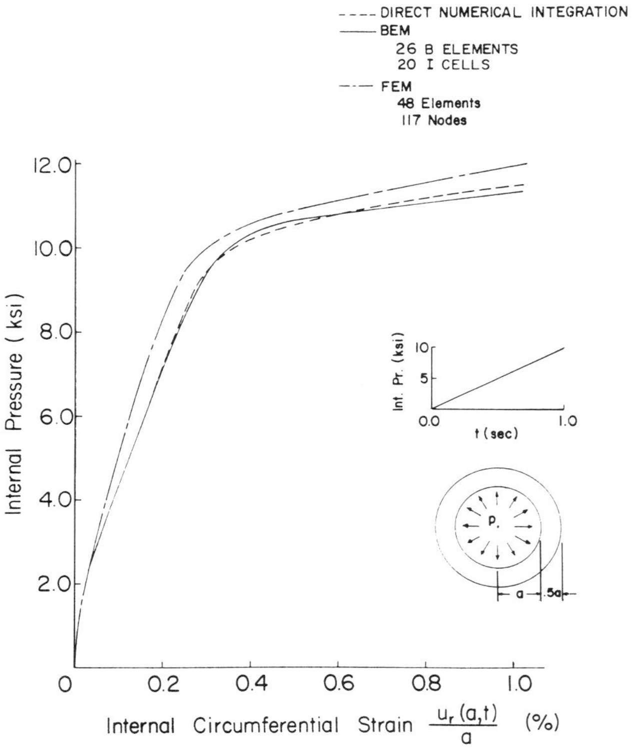

FIG. 5.7. Comparison of direct, BEM and FEM solutions for circular plate with concentric circular cutout under internal pressure increasing at a constant rate. $\dot{p} = 10$ ksi/sec. Annealed 304 SS at 200°C.

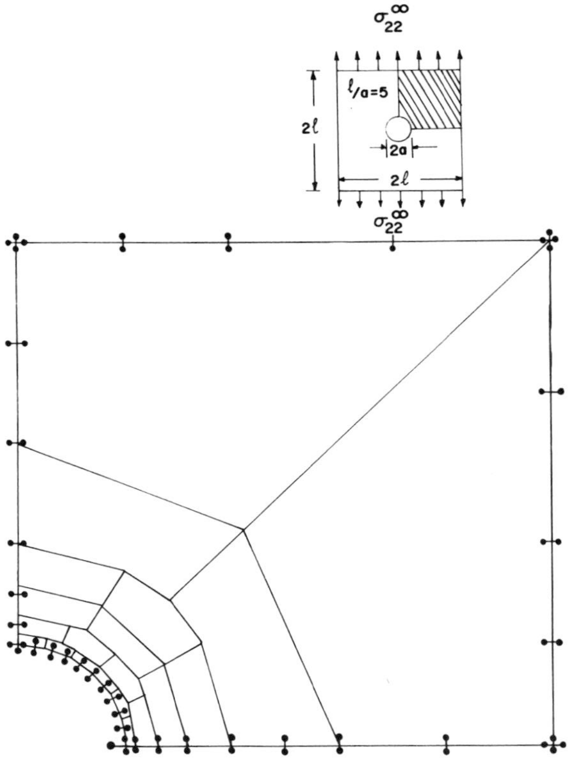

FIG. 5.8. BEM mesh for square plate with circular cutout under uniaxial loading. 37 boundary nodes and 20 internal cells.

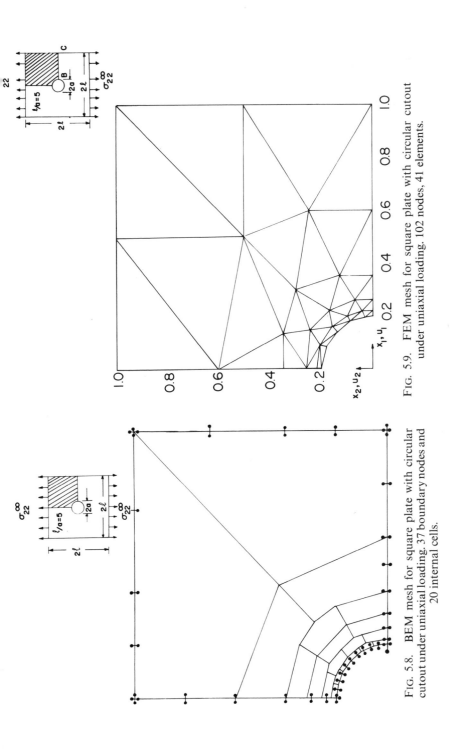

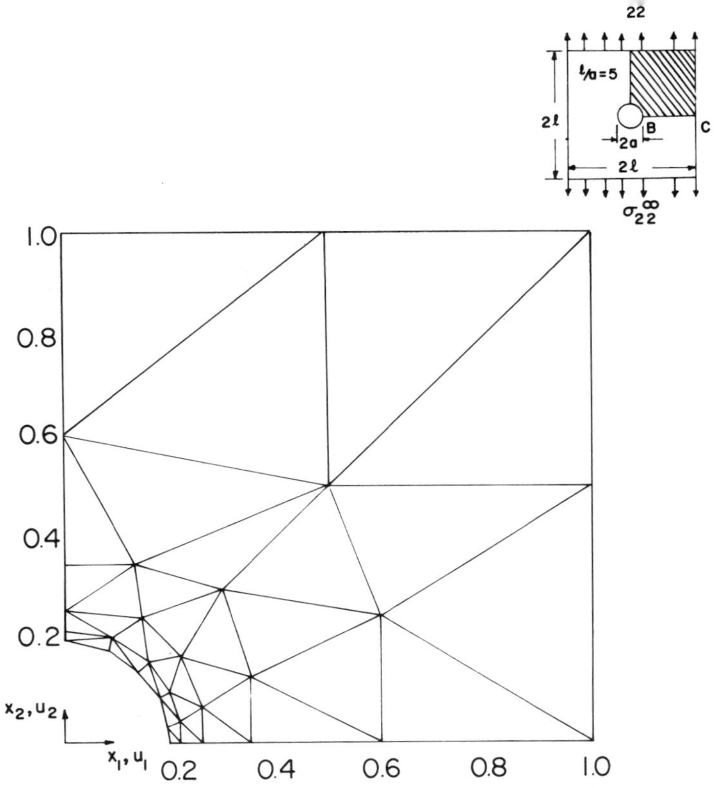

FIG. 5.9. FEM mesh for square plate with circular cutout under uniaxial loading. 102 nodes, 41 elements.

The BEM and FEM results for displacement at the tip of the vertical diameter of the cutout and redistribution of σ_{22} along the line $x_2=0$ are shown in Figs. 5.10 and 5.11 respectively. A direct solution cannot be obtained for this problem. The results obtained from the two methods are in fair agreement. The zero time elastic stress distributions in Fig. 5.11 are within about 3% of the correct solution for the BEM and 6% for the FEM close to the cutout. The solutions are more accurate away from the cutout. The presence of stress concentration near the cutout causes substantial inelastic deformation in this region. This causes rapid unloading of stress close to the cutout while the stress distribution remains nearly unaltered beyond about twice the cutout radius. A vertical equilibrium check on the two long-time solutions reveals that the FEM solution over-predicts the remote stress resultant by about 2% while the BEM solution underpredicts this quantity by about 2·8%. This is considered quite good considering that

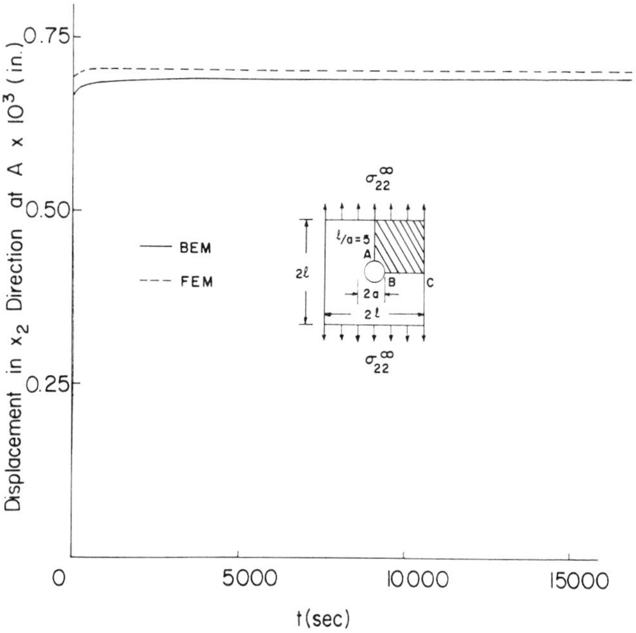

FIG. 5.10. Comparison of BEM and FEM solutions for displacement at A in an annealed 304 SS square plate with circular cutout under uniaxial tension at 400°C. $\sigma_{22}^{\infty}=10$ ksi.

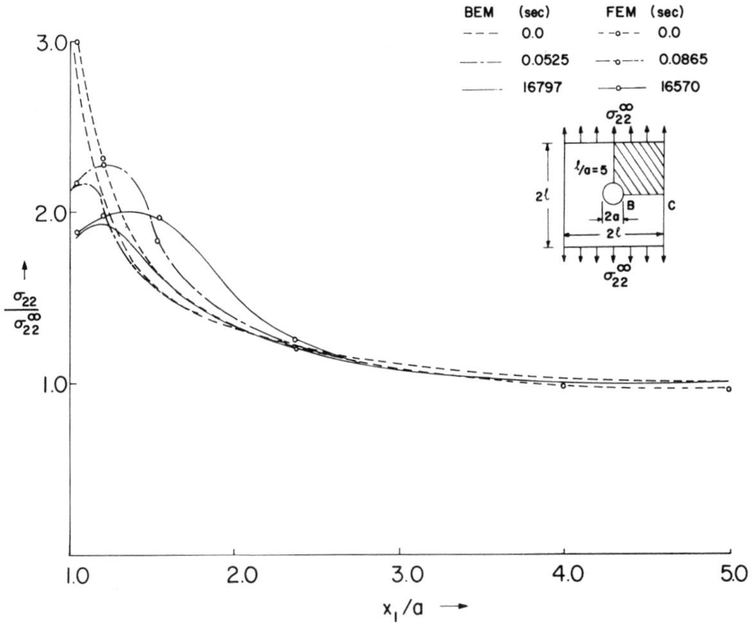

FIG. 5.11. Comparison of BEM and FEM solutions for stress redistribution along BC in an annealed 304 SS square plate with circular cutout under uniaxial tension at 400° C. $\sigma_{22}^{\infty}=10$ ksi.

the BEM solution requires 367 and the FEM 420 time steps to reach the final simulated time. The exact long-time stress distribution, of course, is not known.

5.8.3 Comparison of Computing Times

The CPU times on an IBM 370/168 for these problems are given in Table 5.2. The computing times are seen to be quite reasonable. It is seen that for the problems considered, the BEM program requires proportionately less time compared to FEM as the geometrical complexity of the problem increases. This trend was expected since the number of unknowns in the resultant algebraic system obtained by discretization is proportional to the number of boundary nodes in BEM and to the total number of nodes in FEM. It is expected that the time advantage of the BEM relative to the FEM will become even greater for three-dimensional applications.

TABLE 5.2
BEM AND FEM PROGRAM STATISTICS

Problem	Boundary element method			Finite element method		
	Boundary nodes	Internal elements	CPU sec	Nodes	Elements	CPU sec
1. Uniaxial creep	8	1	7·3	9	2	11·3
2. Uniaxial extension	8	1	3·9	9	2	6·6
3. Circular plate with circular cutout under increasing internal pressure	30	20	63	117	48	127
4. Square plate with circular cutout under constant remote tension	37	20	91	102	41	205

5.9 INTEGRALS OF THE PLANE STRESS KERNELS

The integrals of the kernels U_{ij}, T_{ij}, etc., are evaluated analytically over straight boundary elements and polygonal internal cells. The method used is due to Riccardella[3] (see also reference 6). The local coordinate system and other notation is shown in Figs. 5.12–5.14 for the boundary and area integrals, respectively. The traction and displacement components are piecewise linear on each boundary segment and the non-elastic strain rates are piecewise uniform on each internal cell. Thus, integrals of the type

$$\Delta U_{ij}(p_m, Q_N) = \int_{\Delta c_N} U_{ij}(p_m, Q)\,\mathrm{d}c_Q \tag{5.44}$$

$$\Delta T_{ij}(p_m, Q_N) = \int_{\Delta c_N} T_{ij}(p_m, Q)\,\mathrm{d}c_Q \tag{5.45}$$

$$\Delta \Sigma_{ijk}(p_m, q_n) = \int_{\Delta A_n} \Sigma_{ijk}(p_m, q)\,\mathrm{d}A_q \tag{5.46}$$

$$\Delta V_{ijk}(p_m, Q_N) = \int_{\Delta c_N} V_{ijk}(p_m, Q)\,\mathrm{d}c_Q \tag{5.47}$$

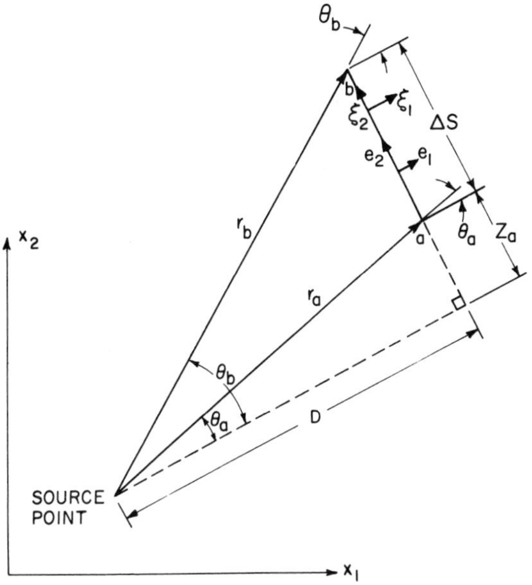

FIG. 5.12. Notation used for the evaluation of boundary integrals.

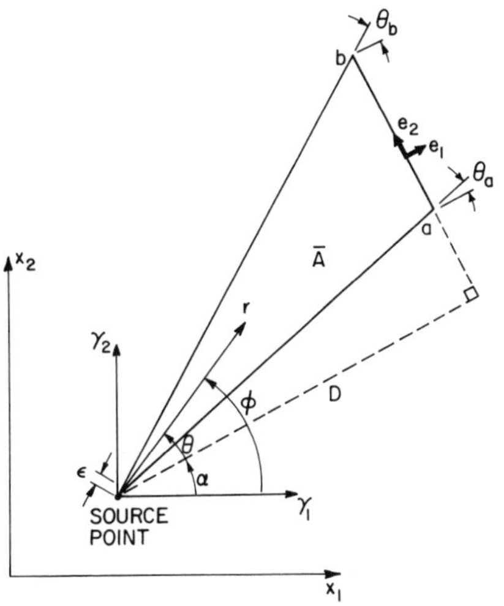

FIG. 5.13. Notation used for the evaluation of area integrals.

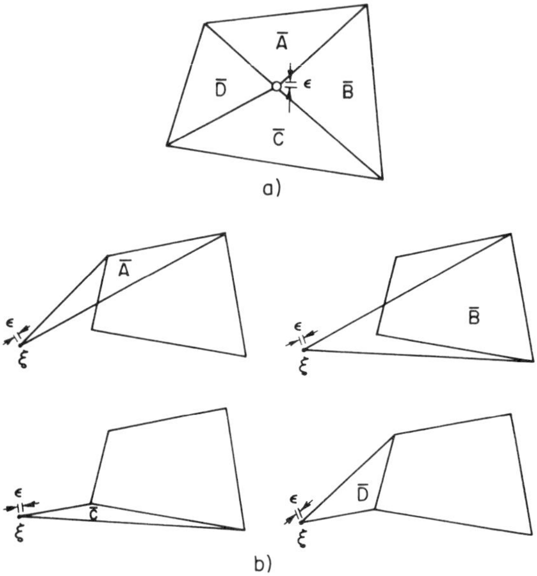

FIG. 5.14. Area integration scheme for polygonal internal cell. (a) Source point inside internal element. (b) Source point outside internal element.

and
$$\Delta T_{ijk}(p_m, Q_N) = \int_{\Delta c_N} T_{ijk}(p_m, Q)\mathrm{d}c_Q \tag{5.48}$$

as well as integrals of the type

$$\Delta \hat{U}_{ij}(p_m, Q_N) = \int_{\Delta c_N} U_{ij}(p_m, Q)c\mathrm{d}c_Q \tag{5.49}$$

(where c is the distance measured along a boundary element) must be determined for the numerical calculations. The integrals of the first kind (ΔU_{ij}, etc.) are given below. The reader is referred to Riccardella[3] for further details of the analytical procedure as well as integrals of the second kind ($\Delta \hat{U}_{ij}$, etc.).

The parameters c_1, c_2, c_3 and c_4 are defined as

$$c_1 = -\frac{1}{8\pi(1-\bar{v})G}, \quad c_2 = (3-4\bar{v})$$

$$c_3 = -\frac{1}{4\pi(1-\bar{v})}, \quad c_4 = (1-2\bar{v})$$

The integrals are

$$\Delta U_{ij}(P_M, Q_N) = c_1 c_2 \delta_{ij} IU_1 - c_1 e_{1i} e_{1j} IU_2$$
$(D \neq 0)$
$$- c_1 (e_{1i} e_{2j} + e_{2i} e_{1j}) IU_3$$
$$- c_1 e_{2i} e_{2j} IU_4$$
$$\Delta U_{ij}(P_M, Q_N) = c_1 c_2 \delta_{ij} IU_5 - c_1 e_{2i} e_{2j} IU_6$$
$(D=0,\ P_M \notin Q_N$ element$)$
$$\Delta U_{ij}(P_M, Q_N) = c_1 c_2 \delta_{ij} IU_7 - c_1 e_{2i} e_{2j} IU_8$$
$(D=0,\ P_M \in Q_N$ element$)$
$$\Delta T_{ij}(P_M, Q_N) = c_3 c_4 \delta_{ij} IT_1 + c_3 e_{1i} e_{1j} IT_2$$
$(D \neq 0)$
$$+ c_3 (e_{1i} e_{2j} + e_{2i} e_{1j}) IT_3$$
$$+ c_3 e_{2i} e_{2j} IT_4 + c_3 c_4 (e_{2i} e_{1j} - e_{1i} e_{2j}) IT_5$$
$$\Delta T_{ij}(P_M, Q_N) = c_3 c_4 (e_{2i} e_{1j} - e_{1i} e_{2j}) IT_5$$
$(D=0)$

For a point p inside the body, $\Delta U_{ij}(p_m, Q_N)$, $\Delta T_{ij}(p_m, Q_N)$ have the same forms as above. In this case, $D=0$, $p_m \in Q_N$ element cannot occur, i.e., the integrals cannot be singular.

Define the tensors (h: integer values 1 or 2)

$$\alpha_{hikl} = e_{hl} \delta_{ik} + e_{hk} \delta_{li} - e_{hi} \delta_{kl}$$
$$\beta_{hikl} = e_{1i} e_{hk} e_{2l} + e_{1k} e_{hl} e_{2i} + e_{1l} e_{hi} e_{2k}$$

Now

$$\Delta\Sigma_{jki}(p_m, \bar{A}_n) = D c_3 c_4 (\alpha_{1ijk} IT_1 + \alpha_{2ijk} IT_5)$$
$$+ D c_3 (e_{1i} e_{1j} e_{1k} IT_2 + \beta_{1ijk} IT_3 + \beta_{2ijk} IT_4 + e_{2i} e_{2j} e_{2k} IV_1)$$

where $\bar{A}_n$ is one of the triangles for the point q_n (see Fig. 5.13). The integral $\Delta\Sigma_{jki}(p_m, q_n)$ is obtained by algebraic summation over all the triangles. The maximum number of sides used for a polygonal internal element is six (see Fig. 5.14).

$$\Delta V_{ijk}(p_m, Q_N) = -c_3 c_4 (\alpha_{1kij} IT_1 + \alpha_{2kij} IT_5)$$
$(D \neq 0)$
$$- c_3 (e_{1i} e_{1j} e_{1k} IT_2 + \beta_{1kij} IT_3 + \beta_{2kij} IT_4 + e_{2i} e_{2j} e_{2k} IV_1)$$
$$\Delta V_{ijk}(p_m, Q_N) = -c_3 (c_4 \alpha_{2kij} + 2 e_{2i} e_{2j} e_{2k}) IT_5$$
$(D=0)$

$$\Delta T_{ijk}(p_m,Q_N)=2Gc_3[-c_4\alpha_{1kij}IT_1-\bar{v}\alpha_{1kij}IT_2-(\bar{v}\alpha_{2kij}+2\bar{v}\beta_{1kij})IT_3$$

$(D\neq 0)$

$$-\bar{v}\beta_{2kij}IT_4+e_{1i}e_{1j}e_{1k}IW_1+\beta_{1kij}IW_2+\beta_{2kij}IW_3$$
$$+e_{2i}e_{2j}e_{2k}IW_4-2\bar{v}e_{1k}\delta_{ij}IT_1-\{(1-\bar{v})e_{1k}\delta_{ij}+e_{1i}e_{1j}e_{1k}\}IT_2$$
$$-\{(1-\bar{v})e_{2k}\delta_{ij}+(1-3\bar{v})(e_{1i}e_{2j}e_{1k}+e_{2i}e_{1j}e_{1k})\}IT_3$$
$$-(1-3\bar{v})e_{2i}e_{2j}e_{1k}IT_4]$$

$$\Delta T_{ijk}(p_m,Q_N)=2Gc_3[c_4\alpha_{1kij}+2\bar{v}\beta_{2kij}+2\bar{v}e_{1k}\delta_{ij}+2(1-3\bar{v})e_{2i}e_{2j}e_{1k}]IW_5$$

$(D=0)$

$$\Delta\Sigma_{ijkl}(p_m,\bar{A}_n)=GC_3[-c_4(e_{1j}\alpha_{1ikl}+e_{1i}\alpha_{1jkl})IT_1-c_4(e_{1j}\alpha_{2ikl}+e_{1i}\alpha_{2jkl})IT_5$$
$$-2(e_{1i}e_{1j}e_{1k}e_{1l})IT_2-(e_{1j}\beta_{1ikl}+e_{1i}\beta_{1jkl})IT_3$$
$$-(e_{1j}\beta_{2ikl}+e_{1i}\beta_{2jkl})IT_4-(e_{1j}e_{2k}e_{2l}e_{2i}+e_{1i}e_{2l}e_{2k}e_{2j})IV_1$$
$$+c_4(e_{1j}\alpha_{1ikl}+e_{1i}\alpha_{1jkl})IS_1-c_4(e_{2j}\alpha_{1ikl}+e_{2i}\alpha_{1jkl})IS_2$$
$$+c_4(e_{1j}\alpha_{2ikl}+e_{1i}\alpha_{2jkl})IS_3-c_4(e_{2j}\alpha_{2ikl}+e_{2i}\alpha_{2jkl})IS_1$$
$$+2(e_{1j}e_{1i}e_{1k}e_{1l})IS_4-(e_{2j}e_{1k}e_{1l}e_{1i}+e_{2i}e_{1k}e_{1l}e_{1j})IS_5$$
$$+(e_{1j}\beta_{1ikl}+e_{1i}\beta_{1jkl})IS_6-(e_{2j}\beta_{1ikl}+e_{2i}\beta_{1jkl})IS_4$$
$$+(e_{1j}\beta_{2ikl}+e_{1i}\beta_{2jkl})IS_7-(e_{2j}\beta_{2ikl}+e_{2i}\beta_{2jkl})IS_6$$
$$+(e_{1j}e_{2k}e_{2l}e_{2i}+e_{1i}e_{2k}e_{2l}e_{2j})IS_8-2(e_{2j}e_{2k}e_{2l}e_{2i})IS_7$$
$$-\left(\frac{2\bar{v}}{1-2\bar{v}}\right)\delta_{ij}\{c_4(2e_{1k}e_{1l}-\delta_{kl})IT_1+c_4(e_{1k}e_{2l}+e_{1l}e_{2k})IT_5$$
$$+(e_{1k}e_{1l})IT_2+(e_{2k}e_{1l}+e_{1k}e_{2l})IT_3$$
$$+(e_{2k}e_{2l})IT_4\}$$
$$+2\bar{v}\delta_{ij}\{2(e_{1k}e_{1l}-e_{2k}e_{2l})IS_1-(e_{2k}e_{1l}+e_{1k}e_{2l})IS_2$$
$$+(e_{1k}e_{2l}+e_{2k}e_{1l})IS_3\}]$$

$\Delta\Sigma_{ijkl}(p_m,q_n)$ is obtained in a manner similar to $\Delta\Sigma_{jki}(p_m,q_n)$.

$$IU_1=D\{\tan\theta_b(\ln r_b-1)-\tan\theta_a(\ln r_a-1)+\theta_b-\theta_a\}$$
$$IU_2=D(\theta_b-\theta_a)$$
$$IU_3=D\ln(r_b/r_a)$$
$$IU_4=D\{(\tan\theta_b-\tan\theta_a)-(\theta_b-\theta_a)\}$$
$$IU_5=S_n\{r_b(\ln r_b-1)-r_a(\ln r_a-1)\}$$ where $S_n=1$ for $\theta=\pi/2$, $S_n=-1$ for $\theta=-\pi/2$

$$IU_6=S_n(r_b-r_a)$$

$$IU_7 = r_b(\ln r_b - 1) + r_a(\ln r_a - 1)$$
$$IU_8 = r_a + r_b$$
$$IT_1 = \theta_b - \theta_a$$
$$IT_2 = (\theta_b - \theta_a) + \sin\theta_b \cos\theta_b - \sin\theta_a \cos\theta_a$$
$$IT_3 = \sin^2\theta_b - \sin^2\theta_a$$
$$IT_4 = (\theta_b - \theta_a) - \sin\theta_b \cos\theta_b + \sin\theta_a \cos\theta_a$$
$$IT_5 = \ln(r_b/r_a)$$
$$IV_1 = \cos^2\theta_b - \cos^2\theta_a + 2\ln(r_b/r_a)$$
$$IW_1 = 3(\theta_b - \theta_a) + 5(\sin\theta_b \cos\theta_b - \sin\theta_a \cos\theta_a) - 2(\sin^3\theta_b \cos\theta_b - \sin^3\theta_a \cos\theta_a)$$
$$IW_2 = 2(\cos^4\theta_a - \cos^4\theta_b)$$
$$IW_3 = (\theta_b - \theta_a) - (\sin\theta_b \cos\theta_b - \sin\theta_a \cos\theta_a) + 2(\sin^3\theta_b \cos\theta_b - \sin^3\theta_a \cos\theta_a)$$
$$IW_4 = 2(\sin^4\theta_b - \sin^4\theta_a)$$
$$IW_5 = S_n\left(\frac{1}{r_b} - \frac{1}{r_a}\right)$$
$$IS_1 = \sin\theta_b \cos\theta_b - \sin\theta_a \cos\theta_a$$
$$IS_2 = \cos^2\theta_b - \cos^2\theta_a$$
$$IS_3 = \sin^2\theta_b - \sin^2\theta_a$$
$$IS_4 = 2(\cos^3\theta_b \sin\theta_b - \cos^3\theta_a \sin\theta_a)$$
$$IS_5 = 2(\cos^4\theta_b - \cos^4\theta_a)$$
$$IS_6 = 2(\sin^2\theta_b \cos^2\theta_b - \sin^2\theta_a \cos^2\theta_a)$$
$$IS_7 = 2(\sin^3\theta_b \cos\theta_b - \sin^3\theta_a \cos\theta_a)$$
$$IS_8 = 2(\sin^4\theta_b - \sin^4\theta_a)$$

REFERENCES

1. MUKHERJEE, S. Corrected boundary-integral equations in planar thermoelastoplasticity. *International Journal of Solids and Structures*, **13**, 331–335 (1977).
2. KUMAR, V. and MUKHERJEE, S. A boundary-integral equation formulation for time-dependent inelastic deformation in metals. *International Journal of Mechanical Sciences*, **19**, 713–724 (1977).
3. RICCARDELLA, P. C. *An Implementation of the Boundary-Integral Technique for Planar Problems of Elasticity and Elasto-Plasticity*. Report No. SM-73-10,

Department of Mechanical Engineering, Carnegie Mellon University, Pittsburg, PA. (1973).

4. MORJARIA, M. and MUKHERJEE, S. Improved boundary-integral equation method for time-dependent inelastic deformation in metals. *International Journal for Numerical Methods in Engineering*, **15**, 97–111 (1980).
5. MENDELSON, A. *Boundary-Integral Methods in Elasticity and Plasticity.* NASA Report No. TND-7418 (1973).
6. MUKHERJEE, S. and KUMAR, V. Numerical analysis of time-dependent inelastic deformation in metallic media using the boundary-integral equation method. American Society of Mechanical Engineers, *Journal of Applied Mechanics*, **45**, 785–790 (1978).
7. MUKHERJEE, S. Thermoviscoplastic response of cylindrical structures using a state variable theory. *Mechanical Behavior of Materials—Proceedings of ICM 3*, Cambridge, England. K. J. Miller and R. F. Smith (eds.), Pergamon Press, Oxford and New York, **2**, 233–242 (1979).
8. MORJARIA, M., SARIHAN, V. and MUKHERJEE, S. Comparison of boundary element and finite element methods in two-dimensional inelastic analysis. *Res Mechanica*, **1**, 3–20 (1980).
9. MORJARIA, M. and MUKHERJEE, S. Finite element analysis of time-dependent inelastic deformation in the presence of transient thermal stresses. *International Journal for Numerical Methods in Engineering*, **17**, 909–921 (1981).
10. MUKHERJEE, S. Time-dependent inelastic deformation of metals by boundary element methods. *Developments in Boundary Element Methods—II*. P. K. Banerjee and R. P. Shaw (eds.), Applied Science Publishers, Barking, Essex, UK, 111–142 (1982).
11. CHAUDONNERET, M. Structure computation in viscoplasticity. Application to two-dimensional calculation of stress concentration. Addendum to *Proceedings for the Second International Symposium on Innovative Numerical Analysis in Applied Engineering Science*, Montreal, Canada. R. P. Shaw *et al.* eds., University of Virginia Press, Charlottesville, Virginia (1980). See also, Methode des equations integrales appliquées à la resolution de problème de viscoplasticite, *Journal de Mécanique Appliqué*, **1**, 113–132 (1977).
12. TELLES, J. C. F. and BREBBIA, C. A. Elastic/viscoplastic problems using boundary elements. *International Journal of Mechanical Sciences.* (In Press).
13. LEMAITRE, J. Sur la determination des lois de comportement des materiaux elastoviscoplastique. Thesis, *ONERA publication No. 135* (1971), also Elastic-viscoplastic constitutive equations for structure quasi-static calculations. Communicated at the Polish Academy of Sciences, Jablonna, Poland (1972).
14. PERZYNA, P. Fundamental problems in viscoplasticity. *Advances in Applied Mechanics*, **9**, 243–377 (1966).
15. KUMAR, V., MUKHERJEE, S., HUANG, F. H. and LI, C-Y. *Deformation in Type 304 Austenitic Stainless Steel*, Electric Power Research Institute Report EPRI NP-1276 for Project 697-1, Palo Alto, CA (1979).
16. GRADSHTEYN, I. S. and RYZHIK, I. W. *Table of Integrals, Series and Products.* Academic Press, New York (1965).

CHAPTER 6

Axisymmetric Viscoplasticity Problems

A boundary element formulation for problems of axisymmetric viscoplastic bodies subjected to axisymmetric mechanical loads is presented in this chapter. While it is fairly straightforward to extend a planar finite element formulation to the axisymmetric case, this is far from true for the boundary element method. The primary reason for the need of considerable effort for the BEM solution is the fact that the axisymmetric kernels contain elliptic functions which cannot be integrated analytically even over boundary elements and internal cells of simple shape. Thus, suitable methods must be developed for the efficient and accurate numerical integration of these singular and sensitive kernels over discrete elements. This aspect of the problem makes the axisymmetric case more difficult to solve than the planar problems discussed in the previous chapter. An important bonus of the development of accurate numerical integration techniques for singular kernels, of course, is that these ideas can then be used in other problems where analytical integration is impossible or impractical because of the need for complicated shapes of boundary elements or internal cells, or for other reasons.

A successful application of the BEM to axisymmetric viscoplastic problems is discussed in this chapter. Numerical results are obtained for several illustrative problems and the BEM results are compared with the results of FEM and direct solutions.

6.1 GOVERNING DIFFERENTIAL EQUATIONS

An axisymmetric body with axisymmetric loading is considered in this chapter. Using polar coordinates R, θ and Z, the nonzero components of displacements, stresses and strains are u_R, u_Z, ε_{RR}, $\varepsilon_{\theta\theta}$, ε_{ZZ}, ε_{RZ} $(=\varepsilon_{ZR})$,

σ_{RR}, $\sigma_{\theta\theta}$, σ_{ZZ} and $\sigma_{RZ}(=\sigma_{ZR})$. All dependent variables are functions of R, Z and t.

The equilibrium equations in rate form, following Timoshenko and Goodyear,[1] in the absence of body forces, are

$$\frac{\partial \dot{\sigma}_{RR}}{\partial R}+\frac{\dot{\sigma}_{RR}-\dot{\sigma}_{\theta\theta}}{R}+\frac{\partial \dot{\sigma}_{RZ}}{\partial Z}=0 \tag{6.1}$$

$$\frac{\partial \dot{\sigma}_{ZR}}{\partial R}+\frac{\dot{\sigma}_{ZR}}{R}+\frac{\partial \dot{\sigma}_{ZZ}}{\partial Z}=0 \tag{6.2}$$

The kinematic equations relating strain and displacement rates are

$$\dot{\varepsilon}_{RR}=\frac{\partial \dot{u}_R}{\partial R},\ \dot{\varepsilon}_{\theta\theta}=\frac{\dot{u}_R}{R} \tag{6.3, 6.4}$$

$$\dot{\varepsilon}_{ZZ}=\frac{\partial \dot{u}_Z}{\partial Z},\ \dot{\varepsilon}_{RZ}=\frac{1}{2}\left(\frac{\partial \dot{u}_R}{\partial Z}+\frac{\partial \dot{u}_Z}{\partial R}\right) \tag{6.5, 6.6}$$

The constitutive equations relating stress and strain rates are

$$\dot{\sigma}_{RR}=\lambda(\dot{\varepsilon}_{RR}+\dot{\varepsilon}_{\theta\theta}+\dot{\varepsilon}_{ZZ})+2G(\dot{\varepsilon}_{RR}-\dot{\varepsilon}^{(n)}_{RR}) \tag{6.7}$$

$$\dot{\sigma}_{\theta\theta}=\lambda(\dot{\varepsilon}_{RR}+\dot{\varepsilon}_{\theta\theta}+\dot{\varepsilon}_{ZZ})+2G(\dot{\varepsilon}_{\theta\theta}-\dot{\varepsilon}^{(n)}_{\theta\theta}) \tag{6.8}$$

$$\dot{\sigma}_{ZZ}=\lambda(\dot{\varepsilon}_{RR}+\dot{\varepsilon}_{\theta\theta}+\dot{\varepsilon}_{ZZ})+2G(\dot{\varepsilon}_{ZZ}-\dot{\varepsilon}^{(n)}_{ZZ})$$

$$\dot{\sigma}_{RZ}=\dot{\sigma}_{ZR}=2G\left[\dot{\varepsilon}_{RZ}-\dot{\varepsilon}^{(n)}_{RZ}\right] \tag{6.9}$$

where λ is the first Lame constant and the equation $\dot{\varepsilon}^{(n)}_{kk}=0$ has been used.

A displacement rate formulation is used for this analysis. Thus, writing eqns. (6.1) and (6.2) in terms of displacement rates yields the Navier equations for displacement rates for this problem. These equations, corresponding to eqn. (3.22) in three dimensions, are

$$\begin{aligned}&\frac{\partial^2 \dot{u}_R}{\partial R^2}+\frac{1}{R}\frac{\partial \dot{u}_R}{\partial R}-\frac{\dot{u}_R}{R^2}+\frac{\partial^2 \dot{u}_R}{\partial Z^2}\\ &\quad+\frac{1}{1-2\nu}\left[\frac{\partial^2 \dot{u}_R}{\partial R^2}+\frac{1}{R}\frac{\partial \dot{u}_R}{\partial R}-\frac{\dot{u}_R}{R^2}+\frac{\partial^2 \dot{u}_Z}{\partial R\partial Z}\right]\\ &\quad=2\left[\frac{\partial \dot{\varepsilon}^{(n)}_{RR}}{\partial R}+\frac{\dot{\varepsilon}^{(n)}_{RR}-\dot{\varepsilon}^{(n)}_{\theta\theta}}{R}+\frac{\partial \dot{\varepsilon}^{(n)}_{RZ}}{\partial Z}\right]\end{aligned} \tag{6.10}$$

$$\frac{\partial^2 \dot{u}_Z}{\partial R^2}+\frac{1}{R}\frac{\partial \dot{u}_Z}{\partial R}+\frac{\partial^2 \dot{u}_Z}{\partial Z^2}$$

$$+\frac{1}{1-2\nu}\left[\frac{\partial^2 \dot{u}_Z}{\partial Z^2}+\frac{1}{R}\frac{\partial \dot{u}_R}{\partial Z}+\frac{\partial^2 \dot{u}_R}{\partial R \partial Z}\right]$$

$$=2\left[\frac{\partial \dot{\varepsilon}^{(n)}_{ZR}}{\partial R}+\frac{\dot{\varepsilon}^{(n)}_{ZR}}{R}+\frac{\partial \dot{\varepsilon}^{(n)}_{ZZ}}{\partial Z}\right] \tag{6.11}$$

The boundary conditions, as usual, require specification of traction or displacement histories on the boundary of the body in the usual way.

6.2 BOUNDARY ELEMENT FORMULATION FOR DISPLACEMENT RATES

A boundary element formulation for displacement rates can be based either on eqn. (3.29) or (3.30) of the three-dimensional case. The coordinate system used is shown in Fig. 6.1. The source point is denoted by $(R,0,Z)$ and the field point by (ρ,θ,ξ). Since the problem is axisymmetric, it is sufficient to choose the source point in the x_1–x_3 plane. The source point coordinates are denoted in this chapter by capital letters. It can, of course, lie inside or on the surface of the body.

The BEM formulation given below is based on eqn. (3.30). An axisymmetric version of the three-dimensional equations can be obtained by integrating the kernels U_{ij}, T_{ij}, etc., for the field point moving around a ring with the source point fixed. This can also be done by formulating a singular body force representation of the ring loads in a direct fashion (see Cruse[2]). The first approach is used here (see reference 3 for related work using fictitious loads).

Integrating eqn. (3.30) for the field point coordinate θ between 0 and 2π results in the equation ($j=1$ and 3, no sum over ρ or ξ)

$$\dot{u}_j(p)=\int_{\partial B}\{U_{\rho j}(p,Q)\dot{t}_\rho(Q)+U_{\xi j}(p,Q)\dot{t}_\xi(Q)$$

$$-T_{\rho j}(p,Q)\dot{u}_\rho(Q)-T_{\xi j}(p,Q)\dot{u}_\xi(Q)\}\,\rho_Q \mathrm{d}c_Q$$

$$+2G\int_B\Big\{U_{\rho j,\rho}(p,q)\dot{\varepsilon}^{(n)}_{\rho\rho}(q)+U_{\rho j,\xi}(p,q)\dot{\varepsilon}^{(n)}_{\rho\xi}(q)$$

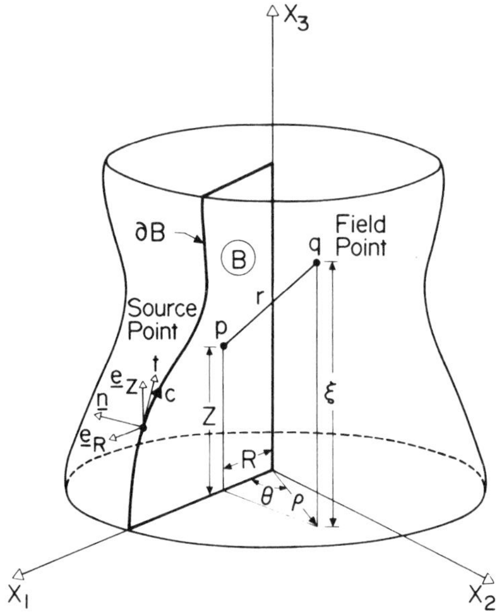

FIG. 6.1. Geometry of the axisymmetric problem.

$$+U_{\xi j,\rho}(p,q)\dot{\varepsilon}^{(n)}_{\xi\rho}(q)+U_{\xi j,\xi}(p,q)\dot{\varepsilon}^{(n)}_{\xi\xi}(q)$$

$$+\frac{U_{\rho j}(p,q)\dot{\varepsilon}^{(n)}_{\theta\theta}(q)}{\rho_q}\Bigg\}\rho_q \mathrm{d}\rho_q \mathrm{d}\xi_q \qquad (6.12)$$

where, because of axisymmetry, $\dot{u}_R(p)=\dot{u}_1(p)$, $\dot{u}_Z(p)=\dot{u}_3(p)$, and $\mathrm{d}c=\sqrt{\mathrm{d}\rho^2+\mathrm{d}\xi^2}$ is an element on the boundary of the $\rho-\xi$ plane. The boundary ∂B in this chapter is that of a R–Z plane and the domain B is a R–Z section of the axisymmetric solid. Thus, only line integrals and area integrals must be evaluated in eqn. (6.12). This reduces the three-dimensional problem to a two-dimensional one.

The kernels $U_{\rho 1}$, etc., are given by the equations[2]

$$U_{\rho 1}(p,q)=\frac{1}{16\pi(1-\nu)G}\frac{k}{\sqrt{R\rho}}\left[\left\{2(3-4\nu)\gamma+\frac{(Z-\xi)^2}{R\rho}\right\}K(k)\right.$$

$$-\left\{2(3-4\nu)(1+\gamma)+\left(\frac{\gamma}{\gamma-1}\right)\frac{(Z-\xi)^2}{R\rho}\right\}E(k)\Bigg] \tag{6.13}$$

$$U_{\xi 1}(p,q)=\frac{(\xi-Z)k}{16\pi(1-\nu)GR\rho\sqrt{R\rho}}\left\{RK(k)+\frac{(\rho-R\gamma)}{\gamma-1}E(k)\right\} \tag{6.14}$$

$$U_{\rho 3}(p,q)=\frac{(Z-\xi)k}{16\pi G(1-\nu)R\rho\sqrt{R\rho}}\left\{\rho K(k)-\frac{(\rho\gamma-R)}{(\gamma-1)}E(k)\right\} \tag{6.15}$$

$$U_{\xi 3}(p,q)=\frac{1}{8\pi(1-\nu)G}\frac{k}{\sqrt{R\rho}}\left\{(3-4\nu)K(k)+\frac{(Z-\xi)^2}{2R\rho(\gamma-1)}E(k)\right\} \tag{6.16}$$

where $K(k)$ and $E(k)$ are complete elliptic integrals of the first and second kind respectively and $k=\sqrt{2/(1+\gamma)}$. Further, $\gamma=1+\{(Z-\xi)^2+(R-\rho)^2\}/2R\rho$.

The traction kernels are given in terms of derivatives of $U_{\rho j}$, etc., by the equations ($j=1$ and 3)

$$\frac{1}{2G}T_{\rho j}(p,Q)=\left\{\left(\frac{1-\nu}{1-2\nu}\right)\frac{\partial U_{\rho j}}{\partial\rho}+\frac{\nu}{(1-2\nu)}\left(\frac{1}{\rho}U_{\rho j}+\frac{\partial U_{\xi j}}{\partial\xi}\right)\right\}n_\rho$$
$$+\frac{1}{2}\left\{\frac{\partial U_{\rho j}}{\partial\xi}+\frac{\partial U_{\xi j}}{\partial\rho}\right\}n_\xi \tag{6.17}$$

$$\frac{1}{2G}T_{\xi j}(p,Q)=\left\{\left(\frac{1-\nu}{1-2\nu}\right)\frac{\partial U_{\xi j}}{\partial\xi}+\left(\frac{\nu}{1-2\nu}\right)\left(\frac{1}{\rho}U_{\rho j}+\frac{\partial U_{\rho j}}{\partial\rho}\right)\right\}n_\xi$$
$$+\frac{1}{2}\left\{\frac{\partial U_{\rho j}}{\partial\xi}+\frac{\partial U_{\xi j}}{\partial\rho}\right\}n_\rho \tag{6.18}$$

where n_ρ and n_ξ are the components of the outward unit normal at the field point on the boundary in the ρ and ξ directions.

The derivatives of the displacement kernels which occur in the above equations as well as in the area integral of eqn. (6.12) are given in reference 4.

The boundary integral equation when the source point p becomes a point P on the boundary ∂B of the body is ($j=1$ and 3, no sum over ρ or ξ)

$$c_{ij}(P)\dot{u}_i(P)=\int_{\partial B}\{U_{\rho j}(P,Q)\dot{\tau}_\rho(Q)+U_{\xi j}(P,Q)\dot{\tau}_\xi(Q)$$

$$- T_{\rho j}(P,Q)\dot{u}_\rho(Q) - T_{\xi j}(P,Q)\dot{u}_\xi(Q)\} \rho_Q \mathrm{d}c_Q$$

$$+2G \int_B \left\{ U_{\rho j,\rho}(P,q)\dot{\varepsilon}^{(\mathrm{n})}_{\rho\rho}(q) + U_{\rho j,\xi}(P,q)\dot{\varepsilon}^{(\mathrm{n})}_{\rho\xi}(q) \right.$$

$$+ U_{\xi j,\rho}(P,q)\dot{\varepsilon}^{(\mathrm{n})}_{\xi\rho}(q) + U_{\xi j,\xi}(P,q)\dot{\varepsilon}^{(\mathrm{n})}_{\xi\xi}(q)$$

$$\left. + \frac{U_{\rho j}(P,q)\dot{\varepsilon}^{(\mathrm{n})}_{\theta\theta}(q)}{\rho_q} \right\} \rho_q \mathrm{d}\rho_q \mathrm{d}\xi_q \tag{6.19}$$

The components of c_{ij} depend, as usual, on the geometry of the boundary at P. These components are determined indirectly by a method analogous to eqn. (3.14). The details of this approach are discussed in Section 6.5.

6.3 STRESS RATES

As discussed in earlier sections, the displacement rate derivatives at source points must first be determined at any time. These may then be used in equations (6.3)–(6.9) to get the stress rates. The derivatives of displacement rates with respect to R and Z at a source point can be determined elementwise numerically or pointwise analytically. Several possible strategies can be used for this purpose. Two strategies have been used to obtain numerical results presented later in the chapter. These are described next.

6.3.1 Numerical Differentiation—the Mixed Method

This is a straightforward approach in which the displacement rates are interpolated within each internal cell by suitable shape functions. These shape functions are then differentiated to obtain the strain rates. This strategy, therefore, uses the boundary element method to determine displacement rates throughout the body and then a method analogous to finite elements to obtain strain rates. It will henceforth be called the mixed method. It has been used by Cathie and Banerjee[5] to solve axisymmetric plasticity problems. This method can be implemented in a simple manner but has the disadvantage of allowing discontinuities in stress rates at internal nodes and across inter-cell boundaries.

6.3.2 Analytical Differentiation—the Strain Rate Gradient Method

This method requires the analytical differentiation of displacement rates at a source point. Two equivalent representations of the displacement rate at an internal point, given by eqns. (3.29) and (3.30) exist. Of these, eqn. (6.12), based on eqn. (3.30), already contains derivatives of U_{ij} in the area integral. These derivatives have singularities of the type $1/r$ so that further differentiation would lead to kernels with singularities of the type $1/r^2$. A method with a free term, as suggested by Bui[6] can possibly be tried to differentiate eqn. (6.12) at a source point, but this still leaves the problem of numerically dealing with strongly singular kernels since analytical integration of these kernels over internal cells is not possible. Equation (3.29), on the other hand, contains U_{ij} which are only logarithmic singular, so that a differentiated version of this equation has singularities of the type $1/r$. These kernels can be integrated numerically as discussed in Section 6.5.3.

The axisymmetric version of eqn. (3.29) is differentiated at a source point to obtain the numerical results presented as 'pure BEM solutions' later in this chapter. This approach gives the equation (see eqn. (3.43))

$$\begin{aligned}\dot{u}_{j,L}(p)=&\int_{\partial B}[U_{\rho j,L}(p,Q)\dot{t}_{\rho}(Q)+U_{\xi j,L}(p,Q)\dot{t}_{\xi}(Q)\\&-T_{\rho j,L}(p,Q)\dot{u}_{\rho}(Q)-T_{\xi j,L}(p,Q)\dot{u}_{\xi}(Q)]\rho_Q \mathrm{d}c_Q\\&+2G\int_{\partial B}[U_{\rho j,L}(p,Q)\{\dot{\varepsilon}^{(\mathrm{n})}_{\rho\rho}(Q)n_{\rho}(Q)+\dot{\varepsilon}^{(\mathrm{n})}_{\rho\xi}(Q)n_{\xi}(Q)\}\\&+U_{\xi j,L}(p,Q)\{\dot{\varepsilon}^{(\mathrm{n})}_{\xi\rho}(Q)n_{\rho}(Q)+\dot{\varepsilon}^{(\mathrm{n})}_{\xi\xi}(Q)n_{\xi}(Q)\}]\rho_Q \mathrm{d}c_Q\\&-2G\int_{B}[U_{\rho j,L}(p,q)\left\{\dot{\varepsilon}^{(\mathrm{n})}_{\rho\rho,\rho}(q)+\frac{\dot{\varepsilon}^{(\mathrm{n})}_{\rho\rho}(q)-\dot{\varepsilon}^{(\mathrm{n})}_{\theta\theta}(q)}{\rho_q}+\dot{\varepsilon}^{(\mathrm{n})}_{\rho\xi,\xi}(q)\right\}\\&+U_{\xi j,L}(p,q)\left\{\dot{\varepsilon}^{(\mathrm{n})}_{\xi\rho,\rho}(q)+\frac{\dot{\varepsilon}^{(\mathrm{n})}_{\xi\rho}(q)}{\rho_q}+\dot{\varepsilon}^{(\mathrm{n})}_{\xi\xi,\xi}(q)\right\}]\rho_q \mathrm{d}\rho_q \mathrm{d}\xi_q\end{aligned} \tag{6.20}$$

where $j=1$ and 3, $L=1$ and 3 and there is no summation over ρ or ξ. Differentiation with respect to a capital letter denotes a source point derivative. By virtue of axisymmetry (see Fig. 6.1)

$$u_{R,R}=u_{1,1},\ u_{R,Z}=u_{1,3},$$

$$u_{Z,R}=u_{3,1},\ u_{Z,Z}=u_{3,3}$$

This method requires accurate values of $\dot{\varepsilon}_{ij}^{(n)}$ on the boundary ∂B, and these can be obtained from the boundary stresses. The boundary stresses can be determined accurately by using the approach discussed in the next section. Equation (6.20) has the drawback of requiring the divergence of the nonelastic strain rates over the domain of B. Numerical realization of these nonelastic strain rate derivatives requires piecewise interpolation of nonelastic strain rates over internal cells. This might be less accurate than direct evaluation of $\dot{\varepsilon}_{ij}^{(n)}$ at internal Gauss points.

It should be noted that kernels in the boundary integrals of stress rate equations never become singular if a source point lies inside the body B.

6.3.3 Boundary Stress Rates

The boundary stress rates are best determined directly from the boundary data. The method outlined below is quite analogous to the strategy discussed for three-dimensional problems in Section 3.3.4. The details, however, are somewhat different.

The boundary integral equation is solved at a given time so that the rates of displacements and tractions are known over the entire boundary. In this discussion, piecewise straight boundary elements are considered on the boundary ∂B. The normal and tangential components of the traction rate vector are first calculated at some point P on ∂B. (P is assumed to lie at a point on ∂B where it is locally smooth.) Now

$$\dot{\sigma}_{nn}=\dot{\tau}_n,\ \dot{\sigma}_{nc}=\dot{\tau}_c$$

where $\dot{\sigma}_{nn}$ and $\dot{\sigma}_{nc}$ are the normal and shearing components of the stress rates at P. As usual, the anticlockwise distance along the boundary element at P is denoted by c (see Fig. 6.1). Next, the normal and tangential components of the displacement rate vector are calculated at P and the tangential derivative of $\dot{u}_c$, $\partial\dot{u}_c/\partial c$, is obtained at P by numerical differentiation along the boundary element. The constitutive equations are written as

$$\frac{\partial \dot{u}_c}{\partial_c}=\dot{\varepsilon}_{cc}=\frac{1}{E}\{\dot{\sigma}_{cc}-\nu(\dot{\sigma}_{nn}+\dot{\sigma}_{\theta\theta})\}+\dot{\varepsilon}_{cc}^{(n)} \tag{6.21}$$

$$\frac{\dot{u}_R}{R}=\dot{\varepsilon}_{\theta\theta}=\frac{1}{E}\{\dot{\sigma}_{\theta\theta}-\nu(\dot{\sigma}_{nn}+\dot{\sigma}_{cc})\}+\dot{\varepsilon}_{\theta\theta}^{(n)} \tag{6.22}$$

The nonelastic strain rates $\dot{\varepsilon}_{RR}^{(n)}$, $\dot{\varepsilon}_{\theta\theta}^{(n)}$, $\dot{\varepsilon}_{ZZ}^{(n)}$ and $\dot{\varepsilon}_{RZ}^{(n)}$ are known at P from the stresses through an appropriate constitutive model. The strain rate $\dot{\varepsilon}_{cc}^{(n)}$ is obtained from these by the usual coordinate transformation

$$\dot{\varepsilon}_{cc}^{(n)}=\dot{\varepsilon}_{RR}^{(n)}n_Z^2+\dot{\varepsilon}_{ZZ}^{(n)}n_R^2-2\dot{\varepsilon}_{RZ}^{(n)}n_Rn_Z \tag{6.23}$$

where $n_R=\mathbf{n}\cdot\mathbf{e}_R$ and $n_Z=\mathbf{n}\cdot\mathbf{e}_Z$.

The eqns. (6.21) and (6.22) are linear equations which can be solved for the unknown stress rates $\dot{\sigma}_{cc}$ and $\dot{\sigma}_{\theta\theta}$. This yields the stress rates $\dot{\sigma}_{nn}$, $\dot{\sigma}_{nc}$, $\dot{\sigma}_{cc}$ and $\dot{\sigma}_{\theta\theta}$. The first three rates can be transformed back to yield the stress rates $\dot{\sigma}_{RR}$, $\dot{\sigma}_{ZZ}$ and $\dot{\sigma}_{RZ}$. Thus

$$\dot{\sigma}_{RR}=\dot{\sigma}_{nn}n_R^2+\dot{\sigma}_{cc}n_Z^2-2\dot{\sigma}_{nc}n_Rn_Z \tag{6.24}$$

$$\dot{\sigma}_{ZZ}=\dot{\sigma}_{nn}+\dot{\sigma}_{cc}-\dot{\sigma}_{RR} \tag{6.25}$$

$$\dot{\sigma}_{RZ}=(\dot{\sigma}_{nn}-\dot{\sigma}_{cc})n_Rn_Z+\dot{\sigma}_{nc}(n_R^2-n_Z^2) \tag{6.26}$$

6.4 FINITE ELEMENT AND DIRECT FORMULATIONS

6.4.1 Finite Element Formulation

The finite element formulation used to obtain the numerical results discussed later in this chapter is identical to that presented for planar problems in Section 5.6 except that axisymmetric (ring) elements are used here. Some numerical results for inelastic deformation in the presence of thermal stresses, using this FEM program, are given in reference 7.

6.4.2 Direct Formulation for Uniform Circular Cylinders in Plane Strain

It is possible to derive a direct solution for the much simpler class of problems of cylinders of uniform circular cross-section in plane strain subjected to axisymmetric pressures.[8] In this case u_Z, ε_{RZ}, ε_{ZR}, σ_{RZ} and σ_{ZR} vanish and the dependent variables are functions of R and t only. Equation (6.10) for the radial displacement rate therefore reduces to

$$\frac{\partial^2\dot{u}_R}{\partial R^2}+\frac{1}{R}\frac{\partial\dot{u}_R}{\partial R}-\frac{\dot{u}_R}{R^2}=\frac{1-2\nu}{1-\nu}\left[\frac{\partial\dot{\varepsilon}_{RR}^{(n)}}{\partial R}+\frac{\dot{\varepsilon}_{RR}^{(n)}-\dot{\varepsilon}_{\theta\theta}^{(n)}}{R}\right] \tag{6.27}$$

and eqn. (6.11) is satisfied identically. This equation can be solved in closed form subject to the boundary conditions

$$\sigma_{RR}(a,t)=-p_i(t),\ \sigma_{RR}(b,t)=-p_o(t)$$

where p_i and p_o are the internal and external pressures at the inside and outside radii, 'a' and 'b' respectively, of the cylinder. The stress rates can then be obtained from the displacement rates.

This problem, for the case of *constant* p_i *and* p_o, has been solved in Section 5.4 as an analytical example of the application of the boundary element method in plane strain. The results for displacement and stress rates are given as eqns. (5.27)–(5.30). For the case of time-dependent loading, the rates of the elastic solutions must be added on to the right-hand sides of eqns. (5.27)–(5.30). These terms, for the stress rates, are:

$$\text{For} \qquad \dot{\sigma}_{RR}: \quad -\dot{p}_i\frac{(b^2-R^2)\,a^2}{(b^2-a^2)\,R^2}-\dot{p}_o\frac{(R^2-a^2)\,b^2}{(b^2-a^2)\,R^2} \tag{6.28}$$

$$\text{For} \qquad \dot{\sigma}_{\theta\theta}: \quad \dot{p}_i\frac{(b^2+R^2)\,a^2}{(b^2-a^2)\,R^2}-\dot{p}_o\frac{(R^2+a^2)\,b^2}{(b^2-a^2)\,R^2} \tag{6.29}$$

$$\text{For} \qquad \dot{\sigma}_{ZZ}: \quad \nu\left[\dot{p}_i\frac{2a^2}{(b^2-a^2)}-\dot{p}_o\frac{2b^2}{(b^2-a^2)}\right] \tag{6.30}$$

A similar term must be added to the right-hand side of eqn. (5.27) for $\dot{u}_R$.

Analogous equations for the case of a circular disc in plane stress can be derived in a similar fashion.

6.5 NUMERICAL IMPLEMENTATION OF BEM

The numerical implementation of the boundary element method and the solution strategy follow the pattern of the three- and two-dimensional problems discussed earlier. An important difference between these problems and those discussed in the previous chapter, however, is that the axisymmetric kernels can no longer be integrated analytically over boundary elements or internal cells. Thus, special methods are required for the accurate and efficient evaluation of these singular integrals. Some of these methods are discussed in this section.

The first step, as usual, is to divide the boundary ∂B of an R–Z section of the cylinder into N_s boundary segments and the interior into n_i internal cells. Denoting by $\dot{u}_i(P_M)$ the components of the displacement rates at a point P which coincides with node M, a discretized version of eqn. (6.19) can be written as (ρ, ξ not summed, $j=1$ and 3)

$$c_{ij}(P_M)\dot{u}_i(P_M)=\sum_{N_s}\int_{\Delta c_N}\{U_{\rho j}(P_M,Q)\dot{\tau}_\rho(Q)+U_{\xi j}(P_M,Q)\dot{\tau}_\xi(Q)$$

$$-T_{\rho j}(P_M,Q)\dot{u}_\rho(Q)-T_{\xi j}(P_M,Q)\dot{u}_\xi(Q)\}\rho_Q \mathrm{d}c_Q$$

$$+2G\sum_{n_i}\int_{\Delta A_n}\Big\{U_{\rho j,\rho}(P_M,q)\dot{\varepsilon}^{(n)}_{\rho\rho}(q)+U_{\rho j,\xi}(P_M,q)\dot{\varepsilon}^{(n)}_{\rho\xi}(q)$$

$$+U_{\xi j,\rho}(P_M,q)\dot{\varepsilon}^{(n)}_{\xi\rho}(q)+U_{\xi j,\xi}(P_M,q)\dot{\varepsilon}^{(n)}_{\xi\xi}(q)$$

$$+\frac{U_{\rho j}(P_M,q)\dot{\varepsilon}^{(n)}_{\theta\theta}(q)}{\rho_q}\Big\}\rho_q \mathrm{d}\rho_q \mathrm{d}\xi_q \tag{6.31}$$

The inclusion of $c_{ij}(P_M)$ means that boundary nodes can be placed at a corner. In the numerical implementation carried out by the author and Sarihan,[9] double nodes are placed at corners to allow for jumps in tractions and normals across corners.

Suitable shape functions are now chosen for the variation of displacement and traction rates along boundary elements and for the variation of nonelastic strain rates over internal cells. This converts eqn. (6.31) into an algebraic system of the type

$$[A]\{\dot{u}\}+[B]\{\dot{\tau}\}=\{\dot{b}\} \tag{6.32}$$

which, as before, must be solved for the unspecified components of the boundary rates of displacements and tractions. Next the displacement rates and finally the stress rates are obtained throughout the body and a march forward time integration scheme is used to obtain the time-histories of the displacements, stresses and strains. The special methods needed for the accurate evaluation of the various terms in eqn. (6.31) and in eqn. (6.20) (if it is used) are discussed next.

6.5.1 Evaluation of Integrals of U_{ij} over Singular Boundary Elements

When a source point P_M lies inside or on the edge of a boundary element Δc_N in eqn. (6.31), this element is termed a singular element. The integrals of U_{ij} and T_{ij} over such elements require special care. The components of the kernel U_{ij} (eqns. (6.13)–(6.16)) have a singularity of the type $\ln r$ when P_M lies in Δc_N and r is the distance between P_M and Q. In such cases, it is fruitful to use the transformation $r=e^2$ which changes $\ln r\,\mathrm{d}r$ to $4e\ln e\,\mathrm{d}e$, which is regular as $e\to 0$.[10] Regular Gaussian integration is performed after this transformation is carried out.

6.5.2 Evaluation of the Tensor c_{ij} and Integrals of T_{ij}

The kernel T_{ij} has a singularity of the type $1/r$ on a singular element. The singular integrals of T_{ij} and the tensor c_{ij} are best determined indirectly by the use of rigid body translation[11] and elastic inflation modes.

Rigid Body Translation in the Z Direction

The elastic problem is considered here. If a rigid body translation $u_Z=1$ is applied to every point of the axisymmetric body, no stresses or tractions are generated in it. In this case, eqn. (6.19) becomes ($j=1$ and 3)

$$c_{3j}(P)=-\int_{\partial B} T_{\xi j}(P,Q)\rho_Q \mathrm{d}c_Q \tag{6.33}$$

If ∂B_c is the part of ∂B which contains P and $\partial\hat{B}$ is the rest,

$$c_{3j}(P)+\int_{\partial B_c} T_{\xi j}(P,Q)\rho_Q \mathrm{d}c_Q = -\int_{\partial\hat{B}} T_{\xi j}(P,Q)\rho_Q \mathrm{d}c_Q \tag{6.34}$$

where the left-hand side of eqn. (6.34) is obtained by numerical evaluation (Gaussian integration) of the nonsingular term on the right.

Inflation Mode in the R Direction

The displacement field $u_R=R$ is a possible solution of the Navier equations for the axisymmetric elasticity problem (see eqn. (6.10)). This gives the tractions

$$\tau_R=2(\lambda+G)n_R,\ \tau_Z=2\lambda n_Z$$

at a point P on the boundary ∂B. If this solution is imposed, eqn. (6.19) becomes

$$c_{1j}(P)R_P=\int_{\partial B}\{2U_{\rho j}(P,Q)(\lambda+G)n_\rho(Q) + 2U_{\xi j}(P,Q)\lambda n_\xi(Q)-T_{\rho j}(P,Q)\rho_Q\}\rho_Q \mathrm{d}c_Q \tag{6.35}$$

which can be written as ($j=1$ and 3, no sum on ρ or ξ)

$$c_{1j}(P)R_P+\int_{\partial B_c} T_{\rho j}(P,Q)\rho_Q^2 \mathrm{d}c_Q = \int_{\partial B}\{2U_{\rho j}(P,Q)(\lambda+G)n_\rho(Q) + 2U_{\xi j}(P,Q)\lambda n_\xi(Q)\}\rho_Q \mathrm{d}c_Q - \int_{\partial\hat{B}} T_{\rho j}(P,Q)\rho_Q^2 \mathrm{d}c_Q \tag{6.36}$$

Once again, the left-hand side of eqn. (6.36) is obtained by numerical evaluation (Gaussian integration) of the right. The length of a boundary element on which ρ varies must be small for this method to be useful.

6.5.3 Evaluation of Area Integral

Terms like $U_{\rho j,\rho}$ in eqn. (6.31) or (6.12) are singular when the source point lies inside or on an internal cell over which the integral is being evaluated. In such a situation, $U_{\rho j,\rho}$ has a $1/r_{pq}$ singularity. The following transformation is useful for the evaluation of the area integral in such cases.

Referring to Fig. 6.2, let $f(R,Z;\ \rho,\xi)$ denote a term like $U_{\rho j,\rho}\dot{\varepsilon}^{(n)}_{\rho\rho}$. The integral to be evaluated over a triangular internal cell is

$$I=\int_{\Delta} f(R,Z;\ \rho,\xi)\mathrm{d}A=\int_{\Delta} f(R,Z;\ \rho,\xi)s\mathrm{d}s\mathrm{d}\psi$$

The coordinates (ρ,ξ) of the field point q are transformed to local polar coordinates (s,ψ). If p coincides with 0, f is singular with a singularity $1/s$. Otherwise, f is regular. Thus $F_1=sf$ is always regular. Writing I in terms of F_1

$$I=\int_{\Delta} F_1(s,\psi)\mathrm{d}s\mathrm{d}\psi$$

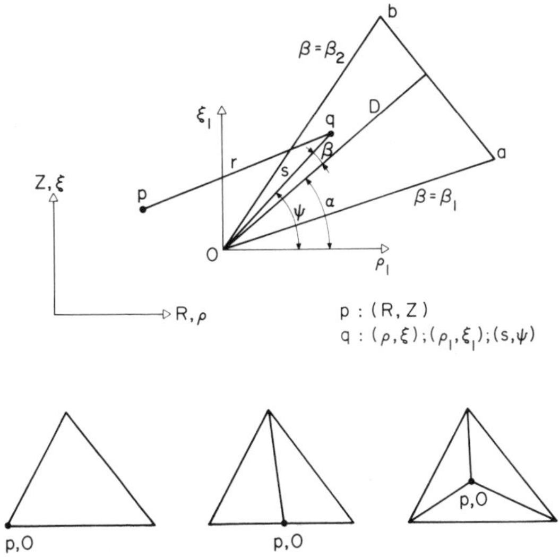

FIG. 6.2. Notation used for the evaluation of area integrals.

(where R, Z are suppressed since the coordinates of the source point are fixed).

First $\beta = \psi - \alpha$ is used to change the integral into

$$I = \int_{\beta_1}^{\beta_2} \int_0^{D/\cos\beta} F_2(s, \beta) \mathrm{d}s \mathrm{d}\beta$$

The following transformation is now used to transform the triangular cell into a unit square

$$s = \frac{\eta D}{\cos\beta}, \quad \beta = (\beta_2 - \beta_1)\phi + \beta_1$$

Using the Jacobian of the transformation,

$$I = \int_0^1 \int_0^1 F_3(\eta, \phi) \frac{D}{\cos\beta} (\beta_2 - \beta_1) \mathrm{d}\eta \mathrm{d}\phi$$

$$= \int_0^1 \int_0^1 F_4(\eta, \phi) \mathrm{d}\eta \mathrm{d}\phi$$

This transformed integral is evaluated by Gaussian quadrature. For the singular cases where the source point does not lie on the vertex of a triangle, the triangle can be broken up into smaller triangles as shown in Fig. 6.2.

The area integral in eqn. (6.20), if it is used, can be evaluated in similar fashion.

It is important to make sure that the integration of singular kernels over triangular internal cells is sufficiently accurate. A check which has proved useful is the numerical evaluation of both sides of the equation

$$\int_\Gamma U_{ij} \dot{\varepsilon}_{ik}^{(n)} n_k \mathrm{d}c = \int_A U_{ij,k} \dot{\varepsilon}_{ik}^{(n)} \mathrm{d}A$$

where Γ is any closed curve enclosing the area A and $\dot{\varepsilon}_{ik}^{(n)}$ is divergence free. A simple choice of $\dot{\varepsilon}_{ik}^{(n)}$ is $\dot{\varepsilon}_{RR}^{(n)} = 1$, $\dot{\varepsilon}_{\theta\theta}^{(n)} = 1$, $\dot{\varepsilon}_{ZZ}^{(n)} = -2$, $\dot{\varepsilon}_{RZ}^{(n)} = 0$.

6.6 NUMERICAL RESULTS—COMPARISON OF VARIOUS SOLUTIONS

6.6.1 Constitutive Model and Parameter Values

The constitutive model of Hart is used to model material behavior of 304 stainless steel at 400 °C. The values of the material parameters used in these calculations are the same as in Section 4.3 except that

$\mathscr{M}=0{\cdot}133\times10^9$ psi. The initial values of σ^* and $\varepsilon_{ij}^{(a)}$ in Hart's model are the same as in Section 4.3.

6.6.2 BEM, FEM and Mixed Methods

The details of the various numerical methods used to get the results reported in this section are given below.

BEM (also called pure BEM). This method uses a piecewise linear description of displacements, tractions and their rates on straight boundary elements. As mentioned before, double nodes are used at corners to allow for jumps in tractions and their rates. The nonelastic strain rates are assumed to be piecewise linear on the boundary elements as well as on triangular internal cells. For the internal cells, the sampling points are placed at the vertices of the triangles. The stress rates are obtained pointwise inside the body from eqn. (6.20) and this requires the evaluation of gradients of nonelastic strain rates at points inside internal cells. Thus, some form of interpolation of these strain rates over internal cells is necessary. The boundary stress rates are obtained from the boundary stress rate algorithm described in Section 6.3.3.

All integrations of kernels are performed numerically. Six or twenty Gauss points are used on the boundary elements. Area integrals are evaluated by the strategy outlined in Section 6.5.3 with a 3×3 grid of Gauss points.

Mixed$_1$. In this case, the displacements and displacement rates throughout the body are obtained by the same approach as described above for the pure BEM method. The stresses and stress rates inside and on the boundary of the body, however, are obtained from a piecewise quadratic interpolation of displacements and displacement rates over triangular internal cells (see Section 6.3.1). The displacement rates are sampled at six points—the vertices and the mid-points of the sides of the triangular cells. There is no longer a need to interpolate nonelastic strain rates over boundary elements. Discretization of nonelastic strain rates over internal cells is carried out as for the pure BEM case. This is not strictly necessary here since nonelastic strain rate gradients need not be computed in this algorithm. However, in view of the strategy used for area integration (Section 6.5.3, see also Fig. 6.2), elimination of the interpolation scheme would require direct evaluation of nonelastic strain rates at many Gauss points for each internal cell. Thus, interpolation appears necessary as an economy measure. The boundary stress rate algorithm (Section 6.3.3) is not used here. Integrations of kernels, required for the evaluation of displacements and their rates, are performed as in the previous case.

Mixed$_2$. This is virtually the same method as Mixed$_1$ with one important difference. Once the stress rates are obtained in B and on ∂B at a given time, the boundary stress rate algorithm is used to recalculate the stress rates at points on ∂B and these values then replace the previously calculated ones. The main idea behind Mixed$_2$ is to smooth out jumps in stresses and stress rates at the boundary nodes. This is done by using central differences to obtain tangential derivatives of displacement rates.

FEM. This is the same program as discussed before for planar problems. A piecewise quadratic interpolation of displacements and their rates is used on axisymmetric finite elements with triangular cross-sections. There is no need for shape functions for the nonelastic strain rates. Instead, these quantities are calculated directly at each time, from the constitutive model, at seven Gauss points which lie inside a triangular element.

6.6.3 Elastic Solutions

Elastic solutions for the stresses in a uniform thick cylinder, in plane strain, subjected to an internal pressure of 10 ksi, have been obtained by various methods. The results, for $b/a = 1{\cdot}5$ ('a' and 'b' are the inside and outside radii of the cylinder) are shown in Table 6.1. The FEM mesh used here is shown in Fig. 6.3. The dots on the same figure are the locations of the boundary nodes for the case B.Nodes = 32. The finer BEM mesh has 20 more boundary nodes placed on the faces Z = constant between the boundary nodes on Fig. 6.3. The BEM for this elastic problem does not require internal discretization.

All the results are seen to be very accurate (maximum deviation from the analytical solution is about $\frac{1}{2}\%$), with the fine mesh BEM delivering practically the exact solution.

6.6.4 Initial Rates for the Viscoplastic Problem

The viscoplastic problem requires step-wise time integration over many time steps in order to obtain the time histories of the quantities of interest. This process is expensive. Thus, it is a very good idea to compare the initial rates of displacements and stresses, as obtained from various methods, for the case of a suddenly applied load. This has been done for the uniform cylinder described above in Section 6.6.3 and the results are tabulated in Tables 6.2 and 6.3. The discretizations used for

TABLE 6.1

ELASTIC SOLUTIONS AT INTERNAL POINTS FOR THE LAME PROBLEM WITH $p_i = 10$ KSI (STRESSES IN PSI)

R/a		*Direct (Lame)*	*FEM (28 elements)*	*BEM B.Nodes = 52*	*BEM B.Nodes = 32*
1·025	σ_{RR}	− 9 133	− 9 138	− 9 132	− 9 125
	$\sigma_{\theta\theta}$	25 133	25 100	25 158	25 232
1·075	σ_{RR}	− 7 576	− 7 574	− 7 576	− 7 576
	$\sigma_{\theta\theta}$	23 576	23 547	23 598	23 664
1·125	σ_{RR}	− 6 222	− 6 205	− 6 224	− 6 229
	$\sigma_{\theta\theta}$	22 222	22 202	22 242	22 302
1·175	σ_{RR}	− 5 038	− 5 007	− 5 040	− 5 047
	$\sigma_{\theta\theta}$	21 038	21 028	21 056	21 111
1·25	σ_{RR}	− 3 520	− 3 508	− 3 523	− 3 530
	$\sigma_{\theta\theta}$	19 520	19 505	19 537	19 586
1·35	σ_{RR}	− 1 877	− 1 873	− 1 879	− 1 886
	$\sigma_{\theta\theta}$	17 877	17 861	17 891	17 935
1·45	σ_{RR}	− 561	− 562	− 563	− 570
	$\sigma_{\theta\theta}$	16 561	16 546	16 574	16 613

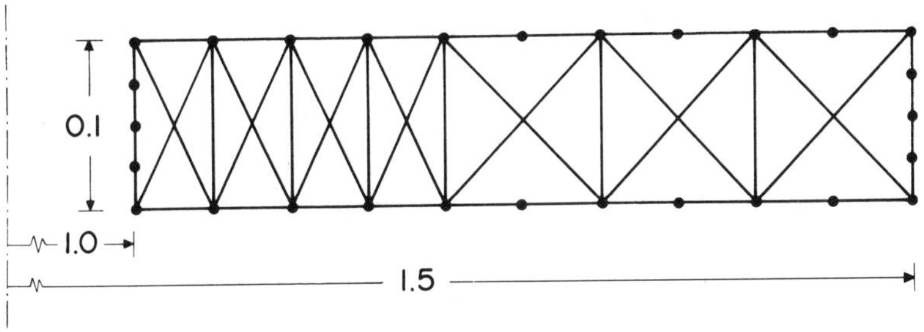

FIG. 6.3. BEM and FEM mesh for uniform cylinder under internal pressure.

BEM: 32 boundary nodes, 28 internal cells
FEM: 73 nodes, 28 elements

the FEM and BEM (with 32 boundary nodes) are shown in Fig. 6.3. The case of BEM with B.Nodes = 52 has the same distribution of internal cells.

The FEM gives the most accurate overall rates in this comparison. As expected, the finer BEM is better than the coarser one. It is felt that there are two main reasons for the difference between BEM and FEM results.

TABLE 6.2

INITIAL RADIAL DISPLACEMENT RATES (IN./SEC) AT INTERNAL POINTS IN A UNIFORM CYLINDER FOR $p_i(0) = 10$ KSI

R/a	*Direct*	*FEM* (28 *elements*)	*BEM* (*B.Nodes* = 52)	*BEM* (*B.Nodes* = 32)
1·025	4 469	4 433	4 666	4 830
1·075	4 078	4 071	4 289	4 434
1·125	3 883	3 845	4 053	4 190
1·175	3 725	3 689	3 889	4 021
1·25	3 556	3 522	3 711	3 837
1·35	3 395	3 362	3 542	3 662
1·45	3 273	3 242	3 415	3 530

TABLE 6.3

INITIAL STRESS RATES (PSI/SEC) FOR SAME CASE AS IN TABLE 6.2.

R		*Direct*	*FEM*	*BEM* *B.Nodes* = 52	*BEM* *B.Nodes* = 32
1·025	$\dot{\sigma}_{RR}$	$-0{\cdot}8418 \times 10^{10}$	$-0{\cdot}1062 \times 10^{11}$	$-0{\cdot}9137 \times 10^{10}$	$-0{\cdot}9587 \times 10^{10}$
	$\dot{\sigma}_{\theta\theta}$	$-0{\cdot}2638 \times 10^{12}$	$-0{\cdot}2633 \times 10^{12}$	$-0{\cdot}2619 \times 10^{12}$	$-0{\cdot}2642 \times 10^{12}$
	$\dot{\sigma}_{ZZ}$	$-0{\cdot}3996 \times 10^{11}$	$-0{\cdot}4084 \times 10^{11}$	$-0{\cdot}4199 \times 10^{11}$	-0.4249×10^{11}
1·075	$\dot{\sigma}_{RR}$	$-0{\cdot}1551 \times 10^{11}$	$-0{\cdot}1639 \times 10^{11}$	$-0{\cdot}1745 \times 10^{11}$	$-0{\cdot}1890 \times 10^{11}$
	$\dot{\sigma}_{\theta\theta}$	$-0{\cdot}8547 \times 10^{11}$	$-0{\cdot}8543 \times 10^{11}$	$-0{\cdot}8304 \times 10^{11}$	$-0{\cdot}8323 \times 10^{11}$
	$\dot{\sigma}_{ZZ}$	$-0{\cdot}8297 \times 10^{10}$	$-0{\cdot}8757 \times 10^{10}$	$-0{\cdot}8550 \times 10^{10}$	$-0{\cdot}8988 \times 10^{10}$
1·125	$\dot{\sigma}_{RR}$	$-0{\cdot}1656 \times 10^{11}$	$-0{\cdot}1684 \times 10^{11}$	$-0{\cdot}1800 \times 10^{11}$	$-0{\cdot}1948 \times 10^{11}$
	$\dot{\sigma}_{\theta\theta}$	$-0{\cdot}4850 \times 10^{10}$	$-0{\cdot}4990 \times 10^{10}$	$-0{\cdot}1996 \times 10^{10}$	$-0{\cdot}1025 \times 10^{10}$
	$\dot{\sigma}_{ZZ}$	$0{\cdot}5543 \times 10^{10}$	$0{\cdot}5304 \times 10^{10}$	$0{\cdot}5959 \times 10^{10}$	$0{\cdot}5881 \times 10^{10}$
1·175	$\dot{\sigma}_{RR}$	$-0{\cdot}1516 \times 10^{11}$	$-0{\cdot}1513 \times 10^{11}$	$-0{\cdot}1623 \times 10^{11}$	$-0{\cdot}1746 \times 10^{11}$
	$\dot{\sigma}_{\theta\theta}$	$0{\cdot}3208 \times 10^{11}$	$0{\cdot}3183 \times 10^{11}$	$0{\cdot}3508 \times 10^{11}$	$0{\cdot}3558 \times 10^{11}$
	$\dot{\sigma}_{ZZ}$	$0{\cdot}1177 \times 10^{11}$	$0{\cdot}1167 \times 10^{11}$	$0{\cdot}1247 \times 10^{11}$	$0{\cdot}1267 \times 10^{11}$
1·25	$\dot{\sigma}_{RR}$	$-0{\cdot}1155 \times 10^{11}$	$-0{\cdot}1170 \times 10^{11}$	$-0{\cdot}1262 \times 10^{11}$	$-0{\cdot}1346 \times 10^{11}$
	$\dot{\sigma}_{\theta\theta}$	$0{\cdot}5324 \times 10^{11}$	$0{\cdot}5260 \times 10^{11}$	$0{\cdot}5597 \times 10^{11}$	$0{\cdot}5796 \times 10^{11}$
	$\dot{\sigma}_{ZZ}$	$0{\cdot}1545 \times 10^{11}$	$0{\cdot}1518 \times 10^{11}$	$0{\cdot}1593 \times 10^{11}$	$0{\cdot}1634 \times 10^{11}$
1·35	$\dot{\sigma}_{RR}$	$-0{\cdot}6458 \times 10^{11}$	$-0{\cdot}6469 \times 10^{10}$	$-0{\cdot}6910 \times 10^{10}$	$-0{\cdot}7387 \times 10^{10}$
	$\dot{\sigma}_{\theta\theta}$	$0{\cdot}5901 \times 10^{11}$	$0{\cdot}5840 \times 10^{11}$	$0{\cdot}6170 \times 10^{11}$	$0{\cdot}6377 \times 10^{11}$
	$\dot{\sigma}_{ZZ}$	$0{\cdot}1683 \times 10^{11}$	$0{\cdot}1664 \times 10^{11}$	$0{\cdot}1753 \times 10^{11}$	$0{\cdot}1804 \times 10^{11}$
1·45	$\dot{\sigma}_{RR}$	$-0{\cdot}1970 \times 10^{10}$	$-0{\cdot}1992 \times 10^{10}$	$-0{\cdot}2110 \times 10^{10}$	$-0{\cdot}2317 \times 10^{10}$
	$\dot{\sigma}_{\theta\theta}$	$0{\cdot}5776 \times 10^{11}$	$0{\cdot}5718 \times 10^{11}$	$0{\cdot}6030 \times 10^{11}$	$0{\cdot}6230 \times 10^{11}$
	$\dot{\sigma}_{ZZ}$	$0{\cdot}1716 \times 10^{11}$	$0{\cdot}1697 \times 10^{11}$	$0{\cdot}1788 \times 10^{11}$	$0{\cdot}1842 \times 10^{11}$

The first is that the BEM uses a piecewise linear representation of displacement rates on the boundary while the FEM has quadratic displacement rate shape functions throughout the body. The second is that the BEM has a piecewise linear representation of nonelastic strain rates over internal cells while the FEM calculates these quantities exactly at Gauss points. In fact, this alone can cause an error of as much as 7% in a nonelastic strain rate component at a point inside an internal cell. It is expected that a higher order interpolation of nonelastic strain rates on internal cells would give rise to a more accurate BEM solution.

6.6.5 Time Histories of Displacements and Stresses for the Viscoplastic Problem

Time histories of displacements for various cases, obtained from different methods, are shown in Figs. 6.4–6.6 and 6.9. The results for axial loading of a uniform cylinder, for increasing and constant loads, respectively, are shown in Figs. 6.4 and 6.5. The results from the direct and FEM calculations coincide in both cases for these simple problems, while the BEM results are very accurate. The $Mixed_2$ approach does relatively poorly for the creep problem in Fig. 6.5.

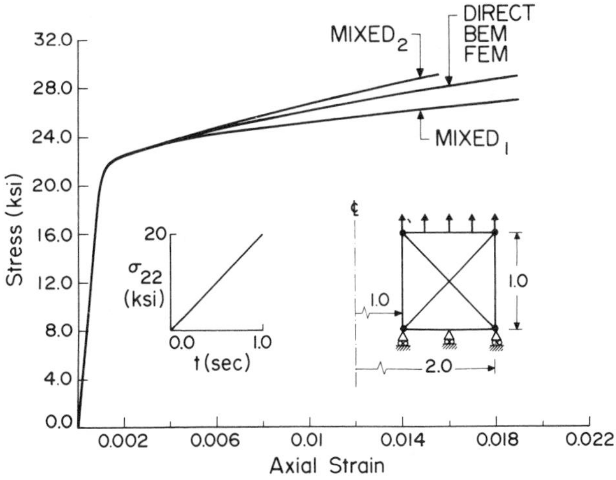

FIG. 6.4. Results for axial loading of a uniform circular cylinder increasing at a constant rate—comparison of various solutions.

BEM: 8 boundary nodes, 4 internal cells
FEM: 12 nodes, 4 elements
$\dot{\sigma}_{22}^{\infty} = 20$ ksi/sec

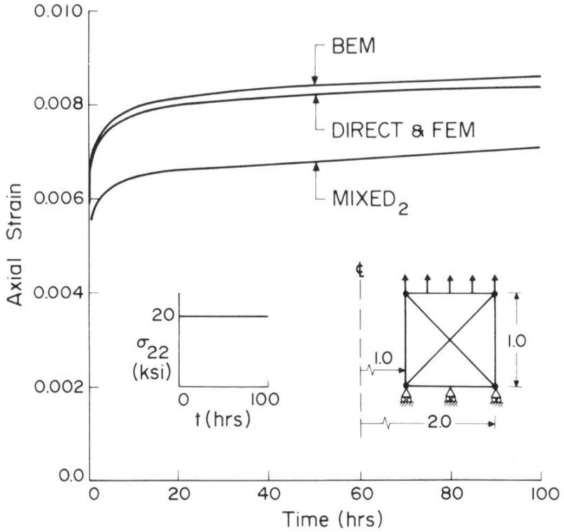

FIG. 6.5. Same situation as Fig. 6.4 but with constant remote axial load. $\sigma_{22}^{\infty} = 20$ ksi

Comparisons for the case of a uniform cylinder subjected to increasing pressure (Fig. 6.6) are very interesting. The discretizations used for the various methods are shown in Fig. 6.3 with the pure BEM and mixed methods using 32 boundary nodes. Once again, the FEM comes out best. As mentioned before, the BEM errors can probably be attributed to the

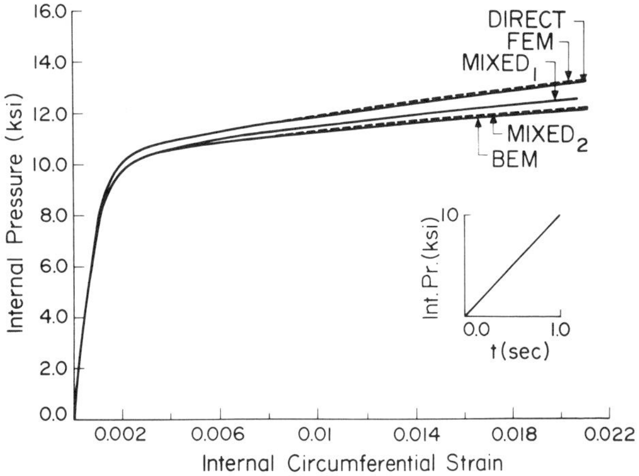

FIG. 6.6. Results for internal pressure on a uniform circular cylinder increasing at a constant rate—comparison of various solutions. $\dot{p} = 10$ ksi/sec. BEM and FEM mesh shown in Fig. 6.3.

boundary discretization used for the displacement rates and the internal discretization for the nonelastic strain rates. Errors from the piecewise linear boundary representation of displacement rates are aggravated in the BEM and Mixed_2 cases which use the boundary stress algorithm, and therefore require numerical differentiation of displacement rates on the boundary ∂B. The Mixed_1 method, which does not use boundary stress rates, is more accurate than BEM and Mixed_2. In any case, the maximum BEM error is around 7% at a simulated strain in excess of 2%. This calculation requires several hundred time steps and the BEM results are considered quite satisfactory for a first attempt at this problem with very complicated kernels.

The redistribution of stresses for the same problem, obtained from the BEM algorithm, are shown in Fig. 6.7. The results show the expected transition from an elastic to an elastic–plastic and finally a plastic stress distribution. The crosses in this figure refer to internal points and the circles to boundary points. The boundary stresses, especially at the inner radius, become progressively less accurate with time. This is attributed to the errors from the boundary stress algorithm as mentioned above. It

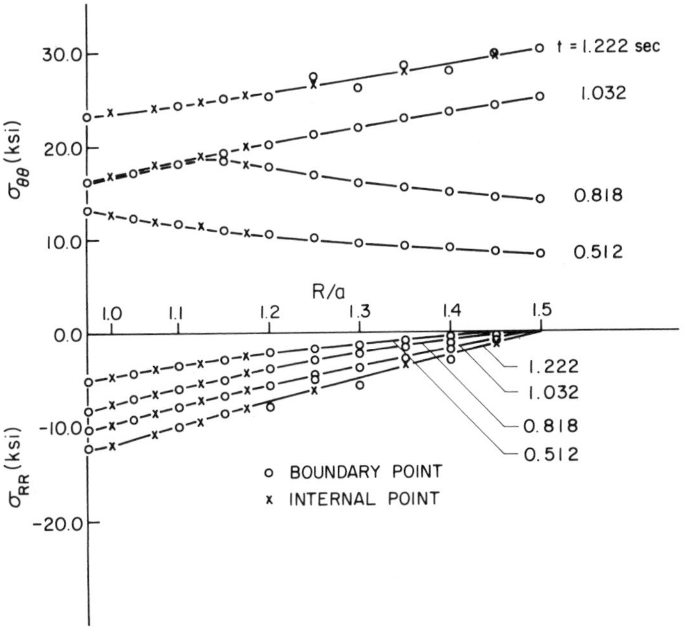

FIG. 6.7. Redistribution of stresses for the loading and geometry of Fig. 6.6. BEM solution.

should be realized that errors in boundary stress rates affect boundary stresses, which, in turn, cause inaccuracies in stress rates and stresses throughout the cylinder as integration proceeds in time.

Sample results for an example of a nonuniform cylinder in plane strain are shown in Fig. 6.9, with the corresponding mesh shown in Fig. 6.8. This is representative of a portion of the core of a Gas Cooled Fast Breeder Reactor Tube (GCFR).[7] The loading is increasing internal pressure. The results for pressure as functions of inside circumferential strain compare well from the BEM and FEM programs. A direct solution is, of course, not possible for this case.

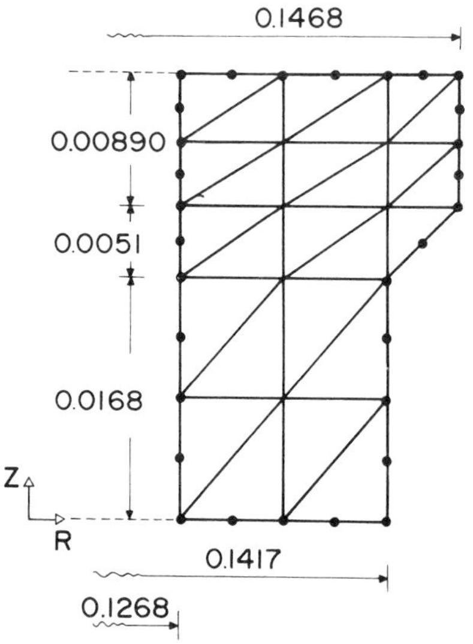

FIG. 6.8. BEM and FEM mesh for GCFR problem.

BEM: 36 boundary nodes, 25 internal cells
FEM: 66 nodes, 25 elements

The pure BEM has the advantage of calculating stresses and stress rates pointwise inside the body. Thus, there are no jumps in these quantities across interelement boundaries. The FEM allows such jumps across elements. The Mixed methods, as implemented here, have internal source points on the boundaries of triangular cells, in order to interpolate, rather than extrapolate, the nonelastic strain rates. This leads to large jumps in stress rates at these internal source points. These rates are directly used (without averaging) in the solution algorithm and the

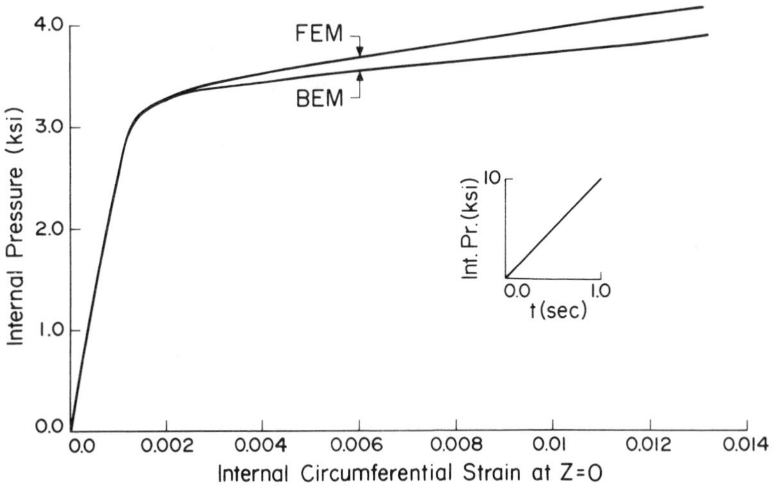

FIG. 6.9. Results for GCFR tube under increasing internal pressure—comparison of BEM and FEM solutions. $\dot{p}=10$ ksi/sec. BEM and FEM mesh shown in Fig. 6.8.

jumps in the rates considerably slow down the stepwise time-integration process as discussed in the next paragraph. Use of source points inside the internal cells would probably improve the performance of the Mixed_1 and Mixed_2 algorithms.

6.6.6 Computer Times

The CPU times on an IBM 370/168 computer, for the various calculations, are given in Table 6.4. The BEM program runs faster for all the internal pressure problems. The mixed method is slow because of large jumps in stress rates at the internal nodes and due to an older inefficient algorithm for evaluating the domain integral. However, the mixed method was considerably slower than the pure BEM even under similar conditions.

6.6.7 Future Directions

It should be emphasized that this is a first BEM attempt at solving this complicated class of problems and the implementation of the BEM can, no doubt, be greatly improved. Possible directions that can be followed are given below.

For the pure BEM:

(a) Use of higher order shape functions for the boundary displace-

TABLE 6.4
COMPUTATIONAL TIME COMPARISON

		BEM	Mixed$_1$[a]	Mixed$_2$[a]	FEM
1. *Uniaxial Tension*					
Boundary Nodes		8	8	8	
Internal Nodes		1	5	5	13
Elements		4	4	4	4
(a) Extension					
CPU time (sec)	*E*	0·727[b]	0·855	0·857	0·112
	T	4·783	23·602	12·536	7·718
(b) Creep					
CPU time (sec)	*E*	0·728[b]	0·88	0·867	0·121
	T	11·654	61·807	36·876	29·172
2. *Internal Pressure*					
Boundary Nodes		32	28	28	
Internal Nodes		7	41	41	73
Elements		28	28	28	28
CPU time (sec)	*E*	7·709[c]	22·972	23·070	0·83
	T	42·495	674·457	614·181	75·046
3. *GCFR tube*					
Boundary Nodes		36			66
Internal Nodes		6			66
Elements		25			25
CPU time (sec)	*E*	8·675[c]			0·468
	T	38·660			44·288

Notes: E = Elastic Time, T = Total Time.
[a]Times are with 20 Gauss points for the elastic solution and an older inefficient algorithm for the domain integral.
[b]Elastic Time with 20 Gauss points.
[c]Elastic Time with 6 Gauss points.
The mixed times have not been modified because they were considerably larger than for the pure BEM even for the same number of Gauss points and similar volume of integral algorithm.

ment rates. A related idea is to retain, say, a piecewise linear boundary representation of displacement rates but to use a higher order boundary interpolation function after the displacement rates have been obtained at the boundary collocation points. This approach does not require too much additional effort but should improve the numerically calculated values of tangential derivatives of displacement rates considerably.

(b) Use of higher order interpolation functions for nonelastic strain rates over internal cells. This appears essential if the BEM accuracy is to be made comparable to the FEM calculations.

For the Mixed methods:

The best, but possibly expensive, suggestion here is to drop the piecewise description of nonelastic strain rates over internal cells completely. Instead, the nonelastic strain rates should be evaluated at internal Gauss points directly from the constitutive model, as is done in the FEM program.

REFERENCES

1. TIMOSHENKO, S. P. and GOODIER, J. N. *Theory of Elasticity*, 3rd. ed., McGraw-Hill (1970).
2. CRUSE, T. A., SNOW, D. W. and WILSON, R. B. Numerical solutions in axisymmetric elasticity. *Computers and Structures*, **7**, 445–451 (1977).
3. KERMANIDIS, T. A numerical solution for axially symmetrical elasticity problems. *International Journal of Solids and Structures*, **11**, 493–500 (1975).
4. SARIHAN, V. PhD Thesis, Department of Theoretical and Applied Mechanics, Cornell University (1982).
5. CATHIE, D. N. and BANERJEE, P. K. Numerical solutions in axisymmetric elastoplasticity. *Innovative Numerical Analysis for the Applied Engineering Sciences*. R. Shaw *et al.* eds. University of Virginia Press, Charlottesville, VA, 331–340 (1980).
6. BUI, H. D. Some remarks about the formulation of three-dimensional thermoelastoplastic problems by integral equations. *International Journal of Solids and Structures*, **14**, 935–939 (1978).
7. MORJARIA, M. and MUKHERJEE, S. Finite element analysis of time-dependent inelastic deformation in the presence of transient thermal stresses. *International Journal for Numerical Methods in Engineering*, **17**, 909–921 (1981).
8. MUKHERJEE, S. Thermoviscoplastic response of cylindrical structures using a state variable theory. *Mechanical Behavior of Materials—Proceedings of ICM 3*, Cambridge, England. K. J. Miller and R. F. Smith (eds.), Pergamon Press, Oxford and New York, **2**, 233–242 (1979).
9. SARIHAN, V. and MUKHERJEE, S. Axisymmetric viscoplastic deformation by the boundary element method, *International Journal of Solids and Structures* (in press).
10. SHIPPY, D. J. Private communication.
11. RIZZO, R. J. and SHIPPY, D. J. An advanced boundary integral equation method for three-dimensional thermoelasticity, *International Journal for Numerical Methods in Engineering*, **11**, 1753–1768 (1977).

CHAPTER 7

Viscoplastic Torsion

The torsion of prismatic shafts of arbitrary cross-section is discussed in this chapter. The shaft material is modelled as elastic–viscoplastic. Numerical solutions for sample problems, obtained by the boundary element and finite element methods, are presented and discussed.

7.1 GOVERNING DIFFERENTIAL EQUATIONS

A prismatic shaft of arbitrary cross-section is twisted by an angle α per unit length of the shaft. The x_3 axis is taken to be parallel to an axis of the shaft so that the x_1 and x_2 axes lie in a plane parallel to a cross-section. The origin is taken in a plane which does not rotate.

The nonzero stress components are σ_{31} $(=\sigma_{13})$ and σ_{32} $(=\sigma_{23})$. A stress function Φ is defined in the usual way such that[1]

$$\sigma_{31}=\frac{\partial \Phi}{\partial x_2}, \quad \sigma_{32}=-\frac{\partial \Phi}{\partial x_1} \tag{7.1}$$

The constitutive equations take the form

$$\dot{\varepsilon}_{31}=\frac{\dot{\sigma}_{31}}{2G}+\dot{\varepsilon}_{31}^{(n)} \tag{7.2}$$

$$\dot{\varepsilon}_{32}=\frac{\dot{\sigma}_{32}}{2G}+\dot{\varepsilon}_{32}^{(n)} \tag{7.3}$$

and the compatibility equation is

$$\frac{\partial \dot{\varepsilon}_{31}}{\partial x_2}-\frac{\partial \dot{\varepsilon}_{32}}{\partial x_1}=-\dot{\alpha} \tag{7.4}$$

Substituting eqns. (7.2) and (7.3) into (7.4) and writing the stresses in terms of the stress function from eqn. (7.1), results in Poisson's equation for the stress function rate in the form

$$\nabla^2 \dot{\Phi} = -2G\dot{\alpha} - 2G\left[\frac{\partial \dot{\varepsilon}_{31}^{(n)}}{\partial x_2} - \frac{\partial \dot{\varepsilon}_{32}^{(n)}}{\partial x_1}\right] \equiv f \tag{7.5}$$

where ∇^2 is the Laplacian operator in two dimensions.

The boundary condition for zero traction on the boundary ∂B of the shaft requires that

$$\Phi = \text{constant on } \partial B. \tag{7.6}$$

If the cross-section is simply connected, this constant can be taken to be zero. If the cross-section is multiply connected, Φ can be taken to be zero on one boundary and its values on the other boundaries must be determined by invoking the requirement of single valued displacements on each boundary.[2]

Finally, use of global equilibrium gives the torque T as

$$\begin{aligned} T &= \int_B (\sigma_{32} x_1 - \sigma_{31} x_2) \mathrm{d}A \\ &= 2 \int_B \Phi \mathrm{d}A \end{aligned} \tag{7.7}$$

where the integral is evaluated over the cross-section B.

The above formulation is in terms of the stress function Φ. Other formulations, in terms of the warping function or the conjugate warping function are possible. Jaswon and Ponter[3] have solved problems of elastic torsion in terms of the warping function, by the boundary element method. Elastic torsion of variable diameter circular shafts has been recently solved by the BEM by Rizzo and Shippy.[4] Mendelson has discussed solution of elasto-plastic torsion problems by the boundary element method.[5]

7.2 BOUNDARY ELEMENT FORMULATIONS

7.2.1 Direct Formulation

Green's theorem states that for two functions u and v

$$\int_B (u\nabla^2 v - v\nabla^2 u) \mathrm{d}A = \int_{\partial B} \left(u\frac{\partial v}{\partial n} - v\frac{\partial u}{\partial n} \right) \mathrm{d}c \tag{7.8}$$

provided that u and v and their partial derivatives up to the required order are continuous in B and on ∂B. Some of these conditions, however, can be relaxed in certain cases and it is sufficient if ∂B is piecewise smooth.[6]

The specific choices for u and v, made here, are $u=\dot{\Phi}$ and $v=\ln r_{pq}$, where the notation of earlier chapters is used. The logarithmic function chosen here is a fundamental (singular) solution of Laplace's equation in two dimensions. It plays the role of the Kelvin's singular solution due to a point load in the elasticity problems discussed earlier. Substituting these into eqn. (7.8) results in the equation

$$\int_B \{\nabla^2(\ln r_{pq})\dot{\Phi}(q)-\ln r_{pq}f(q)\}\,\mathrm{d}A_q$$
$$=\int_{\partial B}\left\{\frac{\partial \ln r_{pQ}}{\partial n_Q}\dot{\Phi}(Q)-\ln r_{pQ}\frac{\partial\dot{\Phi}(Q)}{\partial n_Q}\right\}\mathrm{d}c_Q \tag{7.9}$$

where eqn. (7.5) for $\dot{\Phi}$ has been used. It can be shown by integrating $\nabla^2 \ln r$ on a small circle centered at p, that

$$\nabla^2 \ln r_{pq}=2\pi\Delta(p,q) \tag{7.10}$$

where, as before, Δ is the Dirac-delta function. Using this in eqn. (7.9) results in

$$2\pi\dot{\Phi}(p)=\int_B \ln r_{pq}f(q)\,\mathrm{d}A_q$$
$$+\int_{\partial B}\left[\frac{\partial \ln r_{pQ}}{\partial n_Q}\dot{\Phi}(Q)-\ln r_{pQ}\frac{\partial\dot{\Phi}(Q)}{\partial n_Q}\right]\mathrm{d}c_Q \tag{7.11}$$

At a point P on ∂B where the boundary is locally smooth, it can be proved that

$$\pi\dot{\Phi}(P)=\int_B \ln r_{Pq}f(q)\,\mathrm{d}A_q$$
$$+\int_{\partial B}\left[\frac{\partial \ln r_{PQ}}{\partial n_Q}\dot{\Phi}(Q)-\ln r_{PQ}\frac{\partial\dot{\Phi}(Q)}{\partial n_Q}\right]\mathrm{d}c_Q \tag{7.12}$$

The boundary condition for this problem, for a simply connected region, is $\dot{\Phi}(P)=0$.

7.2.2 Indirect Formulation

The Poisson's equation (7.5) can be transformed into an integral equation by using a single layer potential[5,7]

$$2\pi\dot{\Phi}(p)=\int_{\partial B}\ln r_{pQ}C(Q)\mathrm{d}c_Q+\int_B \ln r_{pq}f(q)\mathrm{d}A_q \tag{7.13}$$

where C is the source strength function (or boundary density function) to be determined from the boundary conditions. This is called an indirect formulation because the boundary integral here involves a function $C(Q)$ which is not directly related to the unknown function $\dot{\Phi}$.

The boundary condition for Φ, eqn. (7.6), at a point P on ∂B where it is locally smooth, gives

$$0=\int_{\partial B}\ln r_{PQ}C(Q)\mathrm{d}c_Q+\int_B \ln r_{Pq}f(q)\mathrm{d}A_q \tag{7.14}$$

where, as before, the boundary integral is interpreted in the sense of a Cauchy principal value.

The stresses at an internal point p are most conveniently obtained by differentiating eqn. (7.13) and using eqn. (7.1). This gives, for $j=1,2$,

$$2\pi\dot{\sigma}_{3j}(p)=\int_{\partial B}H_{3j}(p,Q)C(Q)\mathrm{d}c_Q+\int_B H_{3j}(p,q)f(q)\mathrm{d}A_q \tag{7.15}$$

where

$$H_{31}=\frac{\partial \ln r}{\partial X_2},\quad H_{32}=-\frac{\partial \ln r}{\partial X_1},$$

the derivatives being calculated with respect to the source point p.

Care must be taken to include the residues from the singular kernel when calculating stresses at a boundary point P. It can be shown that

$$2\pi\dot{\sigma}_{3j}(P)=\int_{\partial B}H_{3j}(P,Q)C(Q)\mathrm{d}c_Q+\int_B H_{3j}(P,q)f(q)\mathrm{d}A_q+\pi t_j(P)C(P) \tag{7.16}$$

where ∂B is locally smooth at P and $t_j(P)$ are the components of the unit anticlockwise tangent vector to ∂B at P.

This equation implies that the formulae for the traction rate $\dot{\tau}=\dot{\sigma}_{3j}n_j=\mathrm{d}\dot{\Phi}/\mathrm{d}c_p$ and the tangential stress rate $\dot{\sigma}_{3c}=\dot{\sigma}_{3j}t_j=-\mathrm{d}\dot{\Phi}/\mathrm{d}n_p$ have residues equal to zero and $C(P)/2$ respectively.[3]

7.3 AN ANALYTICAL EXAMPLE

An analytical example for the torsion of a circular shaft is presented next. The coordinate system used is the one in Fig. 5.2. The governing differential equation (7.5), in polar coordinates, is

$$\nabla^2\dot{\Phi} = -2G\dot{\alpha} + \frac{2G}{R}\frac{\partial}{\partial R}(R\dot{\varepsilon}^{(n)}_{z\theta}) = f(q) \tag{7.17}$$

The integral equations for the indirect formulation, (7.13) and (7.14), are now used with $r^2 = R^2 + \rho^2 - 2\rho R\cos\theta$. Equation (7.14) gives

$$\begin{aligned} 0 = Ca\int_0^{2\pi} & K(a; a, \theta)\mathrm{d}\theta \\ & -2G\dot{\alpha}\int_0^a\int_0^{2\pi} K(a; \rho, \theta)\rho\,\mathrm{d}\theta\,\mathrm{d}\rho \\ & +2G\int_0^a\int_0^{2\pi} K(a; \rho, \theta)\frac{\partial}{\partial\rho}\{\rho\dot{\varepsilon}^{(n)}_{z\theta}(\rho)\}\mathrm{d}\theta\,\mathrm{d}\rho \end{aligned} \tag{7.18}$$

where

$$K(R; \rho, \theta) = \ln r = \frac{1}{2}\ln\,(R^2 + \rho^2 - 2\rho R\cos\theta)$$

The integrals of $\ln r$ for the various cases are

$$\int_0^{2\pi} \ln r\,\mathrm{d}\theta = \begin{cases} 2\pi\ln\rho & \rho > R \\ 2\pi\ln R & \rho = R \\ 2\pi\ln R & \rho < R \end{cases} \tag{7.19}$$

Using these integrals results in

$$\begin{aligned} 0 = Ca2\pi\ln a - 2G\dot{\alpha}\int_0^a 2\pi(\ln a)\rho\,\mathrm{d}\rho \\ +2G\int_0^a 2\pi\ln a\frac{\partial}{\partial\rho}\{\rho\dot{\varepsilon}^{(n)}_{z\theta}(\rho)\}\mathrm{d}\rho \end{aligned} \tag{7.20}$$

and solving this

$$C = Ga\dot{\alpha} - 2G\dot{\varepsilon}^{(n)}_{z\theta}(a) \tag{7.21}$$

The corresponding equation for an internal point is

$$2\pi\dot{\Phi}(R)=Ca\int_0^{2\pi}K(R;a,\theta)\mathrm{d}\theta$$
$$-2G\dot{\alpha}\int_0^a\int_0^{2\pi}K(R;\rho,\theta)\rho\mathrm{d}\theta\mathrm{d}\rho$$
$$+2G\int_0^a\int_0^{2\pi}K(R;\rho,\theta)\frac{\partial}{\partial\rho}\{\rho\dot{\varepsilon}_{z\theta}^{(\mathrm{n})}(\rho)\}\mathrm{d}\theta\mathrm{d}\rho \tag{7.22}$$

The double integrals are evaluated by breaking them up into two integrals, one from 0 to R and the other from R to a, and using the appropriate results from eqn. (7.19). This gives

$$2\pi\dot{\Phi}(R)=Ca2\pi\ln a-2G\dot{\alpha}\left\{\pi a^2\ln a+\pi\frac{(R^2-a^2)}{2}\right\}$$
$$+2G(2\pi R\ln R)\dot{\varepsilon}_{z\theta}^{(\mathrm{n})}(R)$$
$$+2G(2\pi a\ln a)\dot{\varepsilon}_{z\theta}^{(\mathrm{n})}(a)$$
$$-2G(2\pi R\ln R)\dot{\varepsilon}_{z\theta}^{(\mathrm{n})}(R)$$
$$-2G\int_R^a 2\pi\dot{\varepsilon}_{z\theta}^{(\mathrm{n})}(\rho)\mathrm{d}\rho \tag{7.23}$$

which, with C from eqn. (7.21), and simplification, gives

$$\dot{\Phi}(R)=\frac{G\dot{\alpha}}{2}(a^2-R^2)-2G\int_R^a\dot{\varepsilon}_{z\theta}^{(\mathrm{n})}(\rho)\mathrm{d}\rho \tag{7.24}$$

Finally,

$$\dot{\sigma}_{z\theta}(R)=-\frac{\mathrm{d}\dot{\Phi}(R)}{\mathrm{d}R}=GR\dot{\alpha}-2G\dot{\varepsilon}_{z\theta}^{(\mathrm{n})}(R) \tag{7.25}$$

The last equation can be easily derived from first principles.

7.4 FINITE ELEMENT FORMULATION

A discretized version of eqn. (7.5), with the boundary conditions

$$\dot{\Phi}=0 \text{ on } \partial B$$
$$\frac{\mathrm{d}\dot{\Phi}}{\mathrm{d}n}=0 \text{ on symmetry lines, if any}$$

(where **n** is normal to a symmetry line) is obtained by a Galerkin type procedure. This is

$$[K]\{\dot{\phi}\}=\{F\} \tag{7.26}$$

where

$$[K]=\sum\int_E [\nabla N]^T[\nabla N]\mathrm{d}A$$

$$\{F\}=-\sum\int_E [N]^T f\mathrm{d}A$$

where ∇ is the gradient operator, E is the element region and the summation is carried out over all the elements. The function f is the same as in eqn. (7.5).

The shape functions $[N]$, are defined in the usual way

$$\dot{\Phi}=[N]\{\dot{\phi}\} \tag{7.27}$$

where $\dot{\phi}$ are the nodal values of the stress function rate.

7.5 NUMERICAL RESULTS

7.5.1 Discretization of Equations

The BEM numerical results presented later in this section are obtained from the indirect formulation discussed in Section 7.2.2. The boundary of the cross-section ∂B is divided into N_s straight boundary elements using N_b ($N_b=N_s$) boundary nodes and the interior of the cross-section is divided into n_i triangular cells. A discretized version of eqn. (7.14) is

$$0=\sum_{N_s}\int_{\Delta c_i} \ln r_{P_M Q} C(Q)\mathrm{d}c_Q$$
$$+\sum_{n_i}\int_{\Delta A_i} \ln r_{P_M q} f(q)\mathrm{d}A_q \tag{7.28}$$

where P_M is a source point which coincides with node M on ∂B, and Δc_i and ΔA_i are boundary internal elements, respectively.

The unknown boundary densities C are assumed to vary linearly over each boundary element with their values to be determined at the nodes which lie at the intersections of these elements. This indirect formulation allows a sampling point to be placed at a corner without any change in eqn. (7.14).

The nonelastic strain rates are interpolated linearly over each triangular cell. Hence, the function f is uniform within each cell. Integrals of $\ln r$ and $c\ln r$ on boundary elements (c is the distance measured from a node along a boundary element) are evaluated analytically when singular and by Gaussian integration (four Gauss points) otherwise. Integrals on triangular internal cells are evaluated by Gaussian integration (seven Gauss points). This method is sufficiently accurate as long as an internal source point lies on the vertex of a triangle, but may not be so for any general location of the source point.

Substitution of the linear functional forms of C into eqn. (7.28) leads to an algebraic system of the type

$$\dot{0}=[A]\{C\}+\{d\} \tag{7.29}$$

The coefficients of the matrix $[A]$ contain boundary integrals of the kernels and the vector $\{d\}$ contains contributions from the area integral. The vector $\{C\}$ contains unknown source strengths at the boundary nodes and the dimension of $\{C\}$ depends only on the number of boundary elements on ∂B.

Equations (7.13) and (7.15) are discretized in similar fashion.

The finite element method solutions are obtained by using triangular finite elements with a quadratic shape function $[N]$. This leads to a linear interpolation of stresses within each element. As in the BEM formulation, the nonelastic strain rates are linearly interpolated within each element. Thus, the function f is uniform within an element.

7.5.2 Constitutive Model

Material behavior is described by an elastic-time-hardening creep constitutive model. The appropriate equations for this model are 2.6, 2.8 and 2.9 of Section 2.1. Here the creep strain $\varepsilon^{(c)}$ is interpreted as the nonelastic strain $\varepsilon^{(n)}$ and eqn. (2.6) is slightly altered to the form

$$\dot{\varepsilon}^{(n)}=\dot{\varepsilon}_c\left(\frac{\sigma}{\sigma_c}\right)^n\mu(\omega+t)^{\mu-1} \tag{7.30}$$

where ω is a small number.

The material parameters used in the numerical calculations are representative of stainless steel at 400° C. These are[8]

$$G=9{\cdot}4\times10^6 \text{ psi}$$
$$\dot{\varepsilon}_c=0{\cdot}277\times10^{-3}\text{ sec}^{-1},\ \sigma_c=0{\cdot}16\times10^6\text{ psi}$$
$$n=7 \qquad \omega=0{\cdot}01\text{ sec}$$
$$\mu=1\text{ (power law creep) or }0{\cdot}5$$

7.5.3 Direct Solution for Shaft with Circular Cross-Section

It is useful to compare the BEM and FEM solutions with a direct solution for a shaft with a circular cross-section. A shaft with a solid circular cross-section of radius 'a' is considered here. Using cylindrical–polar coordinates R, θ and z, as before, the only nonzero stress in this case is $\sigma_{z\theta}$ $(=\sigma_{\theta z})$. Using the usual kinematic assumption that plane sections remain plane,

$$\dot{\varepsilon}_{z\theta}=\frac{R\dot{\alpha}}{2} \tag{7.31}$$

so that

$$\dot{\sigma}_{z\theta}=2G(\dot{\varepsilon}_{z\theta}-\dot{\varepsilon}^{(n)}_{z\theta})=GR\dot{\alpha}-2G\dot{\varepsilon}^{(n)}_{z\theta} \tag{7.32}$$

and the torque is given in terms of the stress by the formula

$$T=\int_0^a 2\pi R^2\sigma_{z\theta}\mathrm{d}R \tag{7.33}$$

The constitutive model gives the following relation for the strain rate in terms of stress in this case

$$\dot{\varepsilon}^{(n)}_{z\theta}=\frac{(3)^{(n+1)/2}}{2}\mu(\omega+t)^{\mu-1}\dot{\varepsilon}_c\left(\frac{\sigma_{z\theta}}{\sigma_c}\right)^n \tag{7.34}$$

7.5.4 Numerical Solution Strategies

The initial nonelastic strains are assumed to be zero. The initial values of the stress function and stresses throughout a cross-section of the shaft and the torque on any section of the shaft are obtained from the corresponding elastic solution. The elastic solution is obtained from the BEM or FEM equations as the case may be. The rates of the nonelastic strains at zero time are obtained from the constitutive equations for the elastic-time-hardening model. The solutions for the two schemes proceed as described in the next two subsections.

Boundary Element Method

The prescribed rate of twist $\dot{\alpha}$ and the initial values of the nonelastic strain rates are used to calculate the vector $\{d\}$ in eqn. (7.29). This equation is solved for the unknown density vector $\{C\}$ on the boundary of a cross-section. Next, $\{C\}$ is used in discretized versions of eqns. (7.13) and (7.15) to get the initial distributions of Φ and the stress rates throughout a cross-section. These rates are used to determine the values of the stress function and stresses after a small time interval Δt. The torque is obtained at this time from eqn. (7.7) using Gaussian integration. This procedure is continued and in this way the time-histories of the relevant variables are obtained. Time integration is carried out by the Euler type stepwise procedure with automatic time-step control described in Section 4.2.1.

Finite Element Method

The prescribed rate of twist $\dot{\alpha}$ and the initial values of $\dot{\varepsilon}_{31}^{(n)}$ and $\dot{\varepsilon}_{32}^{(n)}$ are used to calculate the vector $\{F\}$ in eqn. (7.26). This equation is solved for the initial nodal values of the stress function rate $\dot{\phi}$ and then $\dot{\Phi}$ is obtained throughout a cross-section from eqn. (7.27). The initial stress rates are next obtained from eqn. (7.1), the values of the variables are updated to the time Δt, etc., and the time-histories of the variables are obtained in the same manner as above. The value of the torque at any time is obtained from eqn. (7.7).

Direct Method for Circular Cross-section Shaft

This solution is started by calculating the closed form elastic solution, then using the constitutive equation (7.34) to get $\dot{\varepsilon}_{z\theta}^{(n)}$ at time zero and then eqn. (7.32) to get the initial stress rate. The variables are integrated in time using the same Euler time integration scheme as above. Torque at any time step is obtained from eqn. (7.33).

7.5.5 Comparison of BEM and FEM Results

The BEM and FEM solutions are compared for several illustrative problems. The examples are for solid shafts with circular, square, elliptical and triangular cross-sections respectively. Two time histories of the twist α—the relaxation problem for constant twist and the constant rate of twist problem—are considered for each cross-section. The BEM and FEM results are also compared with the direct solution for the circular cross-section case. Two values of the time-hardening exponent are used in the calculations—$m=1$ (power law creep) and $m=0{\cdot}5$.

Elastic Solutions

Some comparisons among the different solutions for shafts of square and triangular cross-sections are shown in Tables 7.1 and 7.2 respectively. Table 7.1 shows stresses at a boundary and at an inside point in the square cross-section. The normal stress at a boundary point is, of course, supposed to vanish. In general, the results are quite good for the

TABLE 7.1

COMPARISON OF BEM (σ_{ij}^B) AND SERIES (σ_{ij}^S) SOLUTIONS[1] FOR TORSION OF AN ELASTIC SHAFT OF SQUARE CROSS-SECTION (SEE FIG. 7.4)

No. of boundary segments	*No. of internal cells*	$x_1=1\ x_2=0$ $\frac{\sigma_{32}^B-\sigma_{32}^S}{\sigma_{32}^S}\times 100$	$x_1=1\ x_2=0$ $\frac{\sigma_{31}^B}{\sigma_{32}^S}\times 100$	$x_1=0{\cdot}5\ x_2=0{\cdot}5$ $\frac{\sigma_{32}^B-\sigma_{32}^S}{\sigma_{32}^S}\times 100$
4	8	1·61	+11·8	−0·11
8	8	−0·20	+0·14	−0·60
16	8	2·49	1·25	−0·034
8	32	0·08	4·04	−0·092
16	32	−0·16	−0·97	0·033

TABLE 7.2

COMPARISON OF BEM AND FEM AND SERIES[1] SOLUTION FOR TORSION OF AN ELASTIC SHAFT OF TRIANGULAR CROSS-SECTION (FIG. 7.8), $\alpha=0{\cdot}05$ RADIANS

Figure	*Method*		σ_{31}(psi) $x_1=0{\cdot}4\ x_2=0$	ϕ $x_1=0{\cdot}5\ x_2=0{\cdot}519\,64$
7·8	BEM	15 b. segs, 25 cells	$1{\cdot}959\,6\times10^5$	16 774
7·8	BEM	30 b. segs, 100 cells	$1{\cdot}947\,8\times10^5$	16 808
7·8	FEM	25 elements	$1{\cdot}949\,5\times10^5$	16 884
7·8	FEM	100 elements	$1{\cdot}949\,5\times10^5$	16 884
	Series	Ref. 1	$1{\cdot}949\,5\times10^5$	16 884

internal point and improve with mesh refinement for the normal stress at the boundary point. In the nonelastic calculations, the normal stress at a boundary point is forced to be zero and the tangential stress is re-calculated, if necessary, to take this into account. Table 7.2 shows some comparisons for a triangular cross-section. In this case, the results are quite good from the BEM and very good from the FEM calculations.

Nonelastic Solutions

Results for a shaft of circular cross-section are shown in Figs. 7.1–7.3. The BEM and FEM solutions compare very well with the direct solution with the BEM doing slightly better for the constant twisting rate problem ($\dot{\alpha}=0{\cdot}01$ rad/min) in Fig. 7.2. The redistribution of stress in Fig.

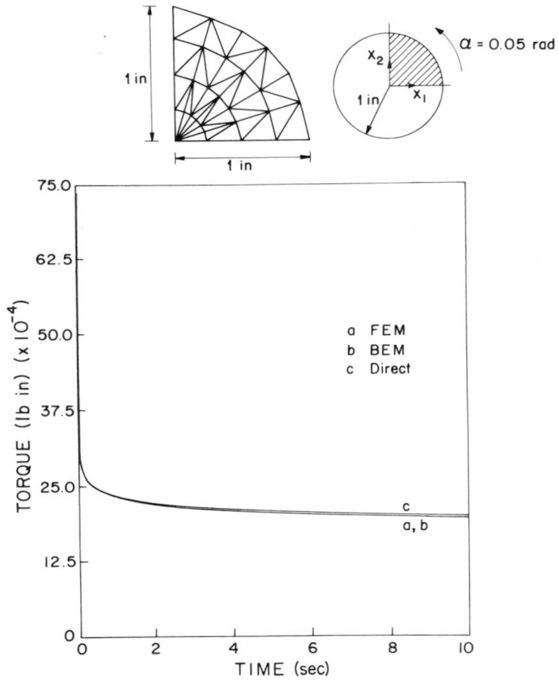

FIG. 7.1. Relaxation of torque in shaft of circular cross-section—comparison of various solutions. Constant $\alpha = 0{\cdot}05$ rad, $\mu = 0{\cdot}5$. BEM mesh: 6 boundary segments, 42 internal cells. FEM mesh: 99 nodes, 42 elements.

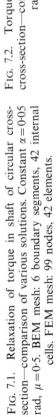

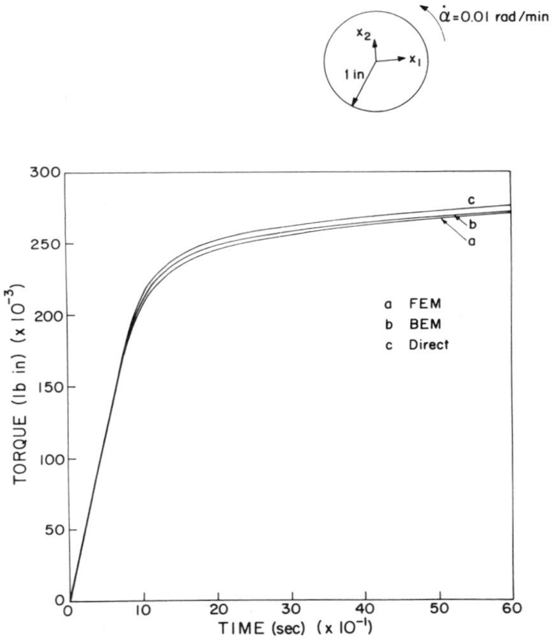

FIG. 7.2. Torque as a function of time in shaft of circular cross-section—comparison of various solutions. Twisting rate $\dot{\alpha} = 0{\cdot}01$ rad/min, $\mu = 0{\cdot}5$.

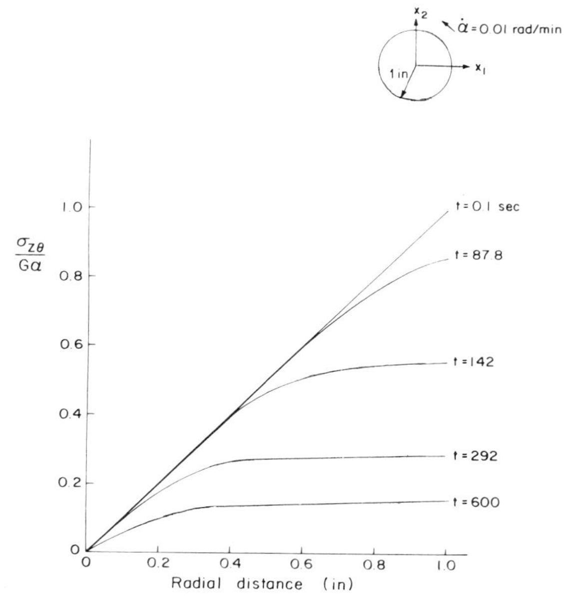

FIG. 7.3. Stress redistribution in a circular shaft under constant rate of twist. BEM solution (within 1% of direct and 5% of FEM solution). $\dot{\alpha}=0{\cdot}01$ rad/min, $\mu=0{\cdot}5$.

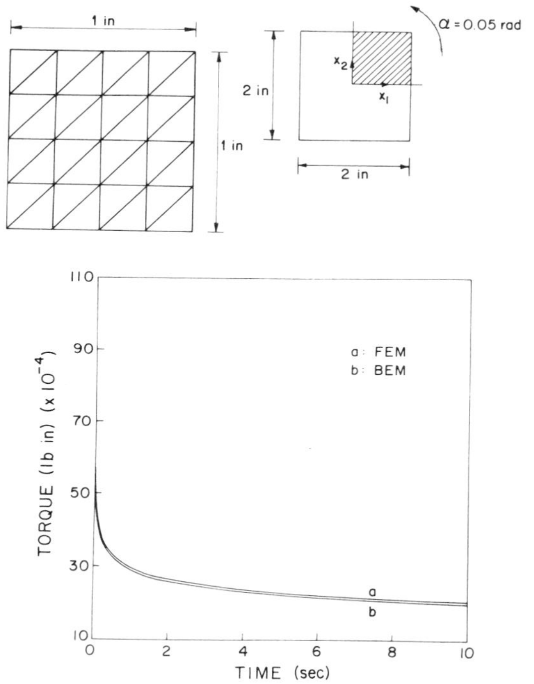

FIG. 7.4. Relaxation of torque in shaft of square cross-section—comparison of BEM and FEM solutions. Constant $\alpha=0{\cdot}05$ rad, $\mu=1{\cdot}0$. BEM mesh: 16 boundary segments, 32 internal cells. FEM mesh: 81 nodes, 32 elements.

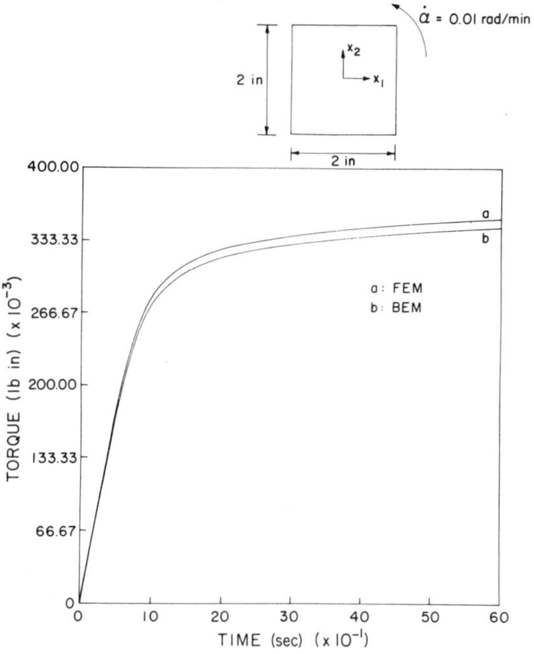

FIG. 7.5. Torque as a function of time in shaft of square cross-section—comparison of BEM and FEM solutions. Twisting rate $\dot{\alpha}=0{\cdot}01$ rad/min, $\mu=0{\cdot}5$.

7.3 shows the expected transition from the elastic to the nonelastic distribution. Again, the BEM solution is very close to the direct solution.

Similar results for shafts with square, elliptical and triangular cross-sections, for the two loading cases α constant and $\dot{\alpha}$ constant, are shown in Figs. 7.4–7.9. Only a quarter of the section needs to be modelled for the square and ellipse. The results for the square and ellipse agree very well for the relaxation problem and quite well for the constant twisting rate problem. For the triangle, however, the FEM results change substantially with mesh refinement and the BEM and FEM results are reasonably close with the refined mesh of Fig. 7.8b.

Computer times

The CPU times on an IBM 370/168 for these problems is shown in Table 7.3. The computing times are of the same order with, generally, the BEM program running somewhat faster than an FEM program with the same internal mesh.

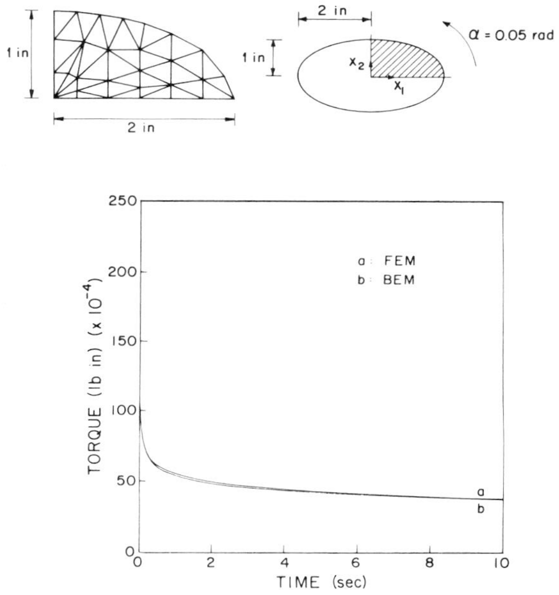

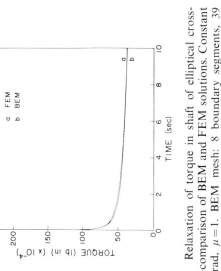

FIG. 7.6. Relaxation of torque in shaft of elliptical cross-section—comparison of BEM and FEM solutions. Constant $\alpha = 0{\cdot}05$ rad, $\mu = 1$. BEM mesh: 8 boundary segments, 39 internal cells. FEM mesh: 95 nodes, 39 elements.

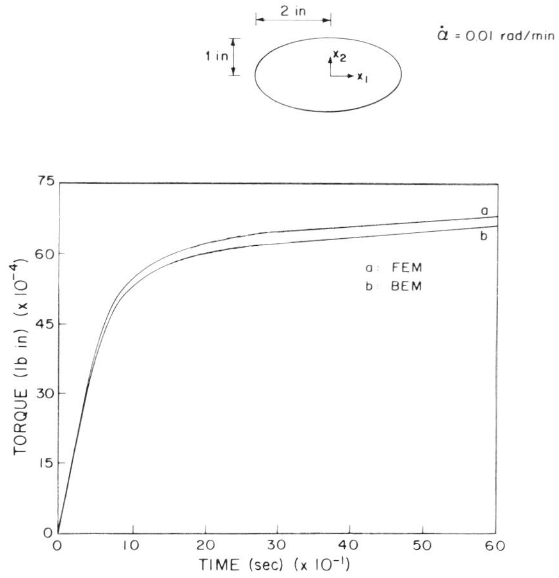

FIG. 7.7. Torque as a function of time in shaft of elliptical cross-section—comparison of BEM and FEM solutions. Twisting rate $\dot{\alpha} = 0{\cdot}01$ rad/min, $\mu = 0{\cdot}5$.

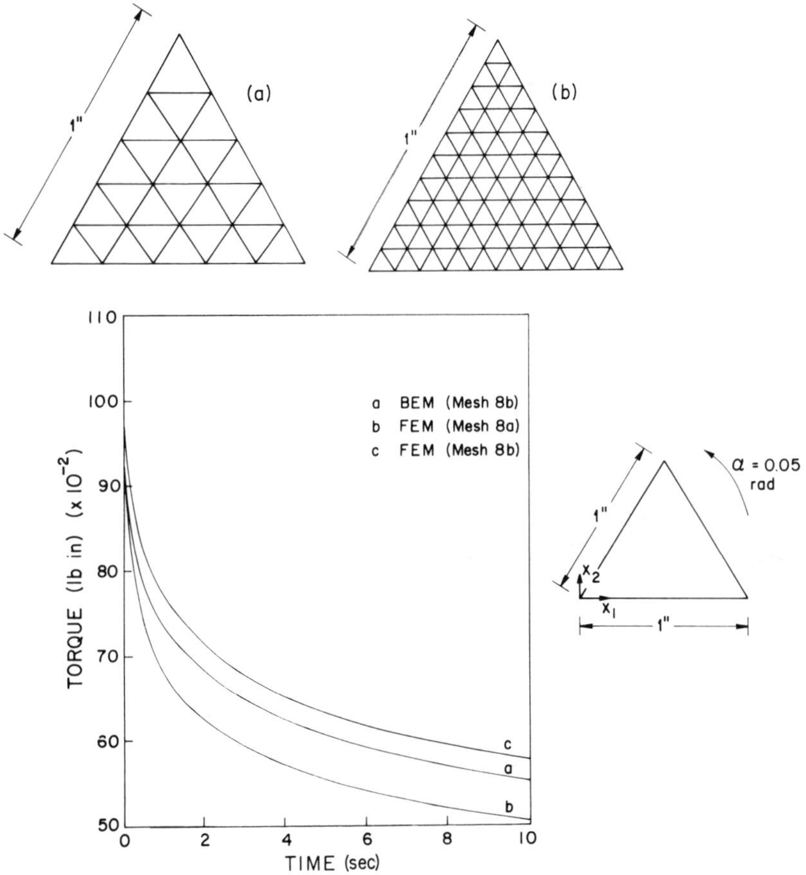

FIG. 7.8. Relaxation of torque in shaft of triangular cross-section—comparison of BEM and FEM solutions. Constant $\alpha = 0{\cdot}05$ rad, $\mu = 1{\cdot}0$. (a) BEM: 15 boundary segments, 25 internal cells. FEM: 66 nodes, 25 elements. (b) BEM: 30 boundary segments, 100 internal cells, FEM: 231 nodes, 100 elements.

TABLE 7.3.
BEM AND FEM PROGRAM STATISTICS

X section	*m*	α *rad*	$\dot{\alpha}$ *rad/min*	*Method*	*Number of time steps*	*CPU Time (secs)*	*Figure*
circular		0·05	0·0	BEM	334	17·73	7·1
				FEM	305	28·90	7·1
	0·5	$\alpha(0)=0$	0·01	BEM	153	9·79	7·2
				FEM	138	13·31	7·2
square	1·0	0·05	0·0	BEM	364	13·63	7·4
				FEM	355	20·71	7·4
	0·5	$\alpha(0)=0$	0·01	BEM	162	9·75	7·5
				FEM	148	9·29	7·5
elliptical	1·0	0·05	0·01	BEM	396	21·01	7·6
				FEM	374	29·03	7·6
	0·5	$\alpha(0)=0$	0·01	BEM	168	10·48	7·7
				FEM	149	12·87	7·7
triangular				BEM	99	24·22	7·8 (Fine Mesh)
	1·0	0·05	0·0	FEM	125	6·26	7·8 (Coarse)
				FEM	104	26·11	7·8 (Fine)
				BEM	141	34·58	7·9 (Fine)
	0·5	$\alpha(0)=0$	0·01	FEM	151	7·42	7·9 (Coarse)
				FEM	134	33·64	7·9 (Fine)

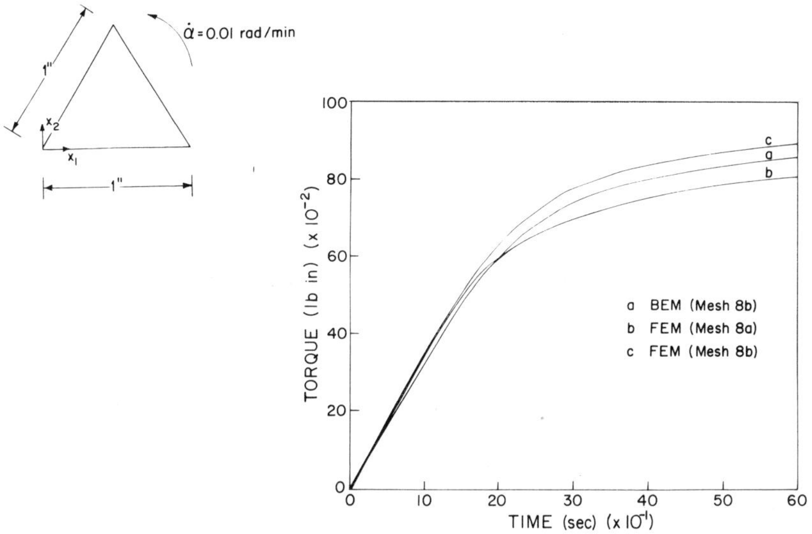

FIG. 7.9. Torque as a function of time in shaft of triangular cross-section—comparison of BEM and FEM solutions. Twisting rate $\dot{\alpha}=0{\cdot}01$ rad/min, $\mu=0{\cdot}5$.

REFERENCES

1. TIMOSHENKO, S. P. and GOODIER, J. N. *Theory of Elasticity*, 3rd ed., McGraw-Hill, New York (1970).
2. SOKOLNIKOFF, I. S. *Mathematical Theory of Elasticity*, 2nd ed., McGraw-Hill, New York (1956).
3. JASWON, M. A. and PONTER, A. R. An integral equation solution of the torsion problem, *Proceedings of the Royal Society*, London, Series A, **273**, 237–246 (1963).
4. RIZZO, F. J. and GUPTA, A. K. A boundary integral equation method for torsion of variable diameter circular shafts and related problems, in *Innovative Numerical Analysis of the Applied Engineering Sciences*, R. Shaw, *et al.* eds., University Press of Virginia, Charlottesville, Virginia, 373–380 (1980).
5. MENDELSON, A. *Boundary Integral Methods in Elasticity and Plasticity*, NASA Report No. TND-7418 (1973).
6. HILDERBRAND, F. B. *Advanced Calculus for Applications*, Prentice Hall, New Jersey (1962).
7. MUKHERJEE, S. and MORJARIA, M. Comparison of boundary element and finite element methods in the inelastic torsion of prismatic shafts. *International Journal for Numerical Methods in Engineering*, **17**, 1576–1588 (1981).
8. ODQUIST, F. K. G. *Mathematical Theory of Creep and Creep Rupture*, Clarendon Press, Oxford (1966).

Stern, Int Jour of Solids + Structures V15 p769
A General BI formulation ...
Song, Mukherjee, Eng Anal, 1986, v1 #3, p86

CHAPTER 8

Bending of Viscoplastic Plates

The transverse deflection of thin, initially flat plates of arbitrary shape is discussed in this chapter. The plate material is modelled as elastic–viscoplastic. Numerical solutions for sample problems, obtained by the direct boundary element method, are presented.

8.1 GOVERNING DIFFERENTIAL EQUATIONS

The analytical formulation of this problem closely follows that for small transverse deflections of thin, laterally loaded elastic plates as presented in the text by Timoshenko and Woinowsky-Krieger.[1] A thin plate of arbitrary shape and uniform thickness h is laterally loaded by a load of intensity q per unit area of the plate. The plate deflections are assumed to be small compared to the thickness of the plate and membrane effects are neglected. The effects of shearing forces and the compressive stress produced by the load q on the plate deformation are neglected. The middle plane of the plate is a plane of symmetry and is assumed to undergo no extension or contraction during bending. It is therefore a neutral surface. The inplane strains in the plate are assumed to remain proportional to the distance from the neutral surface even in the presence of inelastic deformation. The formulation is carried out in terms of the time rates of change of the relevant variables.

The middle plane of the undeformed plate is the x_1x_2 plane and w is the deflection of the middle plane in the x_3 direction. Since the plate deflections and slopes are small, the curvature rates $\dot{\kappa}_1$ and $\dot{\kappa}_2$ and the rate of twist of the middle surface with respect to the x_1 and x_2 axes, $\dot{\kappa}_{12}$, are related to the deflection rate by the following equations

$$\dot{\kappa}_1 = -\frac{\partial^2 \dot{w}}{\partial x_1^2},\ \dot{\kappa}_2 = -\frac{\partial^2 \dot{w}}{\partial x_2^2},\ \dot{\kappa}_{12} = \frac{\partial^2 \dot{w}}{\partial x_1 \partial x_2} \tag{8.1}$$

The strain rates are, by assumption

$$\begin{aligned} \dot{\varepsilon}_{11} &= \dot{\varepsilon}_{11}^{(e)} + \dot{\varepsilon}_{11}^{(n)} = x_3 \dot{\kappa}_1 \\ \dot{\varepsilon}_{22} &= \dot{\varepsilon}_{22}^{(e)} + \dot{\varepsilon}_{22}^{(n)} = x_3 \dot{\kappa}_2 \\ \dot{\varepsilon}_{12} &= \dot{\varepsilon}_{12}^{(e)} + \dot{\varepsilon}_{12}^{(n)} = -x_3 \dot{\kappa}_{12} \end{aligned} \tag{8.2}$$

and the stress rates, from Hooke's Law, become

$$\begin{aligned} \dot{\sigma}_{11} &= \frac{E}{1-\nu^2}(\dot{\varepsilon}_{11}^{(e)} + \nu \dot{\varepsilon}_{22}^{(e)}) \\ &= \frac{E x_3}{1-\nu^2}(\dot{\kappa}_1 + \nu \dot{\kappa}_2) - \frac{E}{1-\nu^2}(\dot{\varepsilon}_{11}^{(n)} + \nu \dot{\varepsilon}_{22}^{(n)}) \\ \dot{\sigma}_{22} &= \frac{E x_3}{1-\nu^2}(\dot{\kappa}_2 + \nu \dot{\kappa}_1) - \frac{E}{1-\nu^2}(\dot{\varepsilon}_{22}^{(n)} + \nu \dot{\varepsilon}_{11}^{(n)}) \\ \dot{\sigma}_{12} &= 2G\dot{\varepsilon}_{12}^{(e)} = -2G x_3 \dot{\kappa}_{12} - 2G\dot{\varepsilon}_{12}^{(n)} \end{aligned} \tag{8.3}$$

The rates of the bending moments $\dot{M}_1$ and $\dot{M}_2$ and the twisting moment $\dot{M}_{12}$ are

$$\dot{M}_1 = \int_{-h/2}^{h/2} \dot{\sigma}_{11} x_3 \mathrm{d}x_3,\ \dot{M}_2 = \int_{-h/2}^{h/2} \dot{\sigma}_{22} x_3 \mathrm{d}x_3,$$

$$\dot{M}_{12} = -\int_{-h/2}^{h/2} \dot{\sigma}_{12} x_3 \mathrm{d}x_3 \tag{8.4}$$

The equilibrium equation of a plate element is[1]

$$\frac{\partial^2 \dot{M}_{11}}{\partial x_1^2} + \frac{\partial^2 \dot{M}_{22}}{\partial x_2^2} - \frac{2\partial^2 \dot{M}_{12}}{\partial x_1 \partial x_2} = -\dot{q} \tag{8.5}$$

Substituting eqns. (8.1), (8.3) and (8.4) into (8.5), the governing differential equation for the deflection rate is obtained as the nonhomogeneous biharmonic equation

$$\nabla^4 \dot{w} = \frac{\dot{q}}{D} - \frac{12}{h^3} f^{(n)}(x_1, x_2, t) \equiv g(x_1, x_2, t) \tag{8.6}$$

where $\nabla^4 = \frac{\partial^4}{\partial x_1^4} + 2\frac{\partial^4}{\partial x_1^2 \partial x_2^2} + \frac{\partial^4}{\partial x_2^4}$,

$$f^{(n)} = \frac{\partial^2}{\partial x_1^2}\int_{-h/2}^{h/2} (\dot{\varepsilon}_{11}^{(n)} + \nu\dot{\varepsilon}_{22}^{(n)})x_3 \mathrm{d}x_3 + \frac{\partial^2}{\partial x_2^2}\int_{-h/2}^{h/2} (\dot{\varepsilon}_{22}^{(n)} + \nu\dot{\varepsilon}_{11}^{(n)})x_3 \mathrm{d}x_3$$

$$+ 2(1-\nu)\frac{\partial^2}{\partial x_1 \partial x_2}\int_{-h/2}^{h/2} \dot{\varepsilon}_{12}^{(n)} x_3 \mathrm{d}x_3 \tag{8.7}$$

and $D = Eh^3/\{12(1-\nu^2)\}$.

Some simple boundary conditions on the boundary ∂B of the plate are:

$$\text{clamped plate: } \dot{w} = 0, \frac{\partial \dot{w}}{\partial \mathbf{n}} = 0 \tag{8.8}$$

$$\text{simple supported plate: } \dot{w} = 0,\ \dot{M}_n = 0 \tag{8.9}$$

where $\mathbf{n}$ is the unit outward normal to ∂B and M_n is the moment obtained by integrating $\dot{\sigma}_{nn}x_3$ through the plate thickness as in eqns. (8.4). On a straight edge of a plate, the last of eqns. (8.9) leads to (see Section 8.6.3)

$$\nabla^2 \dot{w} = 0 \tag{8.10}$$

8.2 BOUNDARY ELEMENT FORMULATIONS

Several integral formulations for the plate bending problem are possible. An indirect formulation appears in reference 2 where bending of elastic plates is discussed. Recent formulations for the elastic problem directly using physical variables like moment and shear on the boundary are given in references 3 and 4. The direct formulation for the inelastic problem, given next, follows Rzasnicki and Mendelson[5] where the biharmonic equation for the stress function for a plane elastoplastic problem is discussed.

8.2.1 Direct Formulation

This formulation is based on Green's theorem, eqn. (7.8). Writing $u = \nabla^2 \phi$, eqn. (7.8) becomes

$$\int_B \{\nabla^2\phi\nabla^2 v - v\nabla^4\phi\}\mathrm{d}A = \int_{\partial B}\left\{\nabla^2\phi\frac{\partial v}{\partial n} - v\frac{\partial}{\partial n}(\nabla^2\phi)\right\}\mathrm{d}c \tag{8.11}$$

A similar equation can be written by interchanging ϕ and v in the above equation. Subtracting this equation from eqn. (8.11) results in

$$\int_B \{(\nabla^4 v)\phi - v\nabla^4\phi\}\mathrm{d}A$$
$$= \int_{\partial B}\left\{\frac{\partial}{\partial n}(\nabla^2 v)\phi - (\nabla^2 v)\frac{\partial\phi}{\partial n} + \frac{\partial v}{\partial n}(\nabla^2\phi)\right.$$
$$\left. - v\frac{\partial}{\partial n}(\nabla^2\phi)\right\}\mathrm{d}c \qquad (8.12)$$

The specific choices made here are $\phi = \dot{w}$ and $v = r^2 \ln r$ where r, as usual, is the distance between a source point and a field point. The function $\mathrm{r}^2 \ln r$ is a fundamental solution of the biharmonic equation. Making these substitutions into eqn. (8.12) and using eqn. (8.6) results in

$$\int_B \{\nabla^4(r^2\ln r)\dot{w} - (r^2\ln r)g\}\mathrm{d}A$$
$$= \int_{\partial B}\left[\frac{\partial}{\partial n}\{\nabla^2(r^2\ln r)\}\dot{w} - \nabla^2(r^2\ln r)\frac{\partial\dot{w}}{\partial n}\right.$$
$$\left. + \frac{\partial}{\partial n}(r^2\ln r)\nabla^2\dot{w} - \mathrm{r}^2\ln r\frac{\partial}{\partial n}(\nabla^2\dot{w})\right]\mathrm{d}c \qquad (8.13)$$

By integrating $\nabla^2(r^2\ln r)$ on a small circle centered at the source point p, it can be demonstrated that

$$\nabla^4(r^2\ln r) = 8\pi\Delta(p,q) \qquad (8.14)$$

so that $\int_B \nabla^4(r^2\ln r)_{pq}\dot{w}(q)\mathrm{d}A_q = 8\pi\dot{w}(p)$, and, writing eqn. (8.13) in explicit notation[6]

$$8\pi\dot{w}(p) - \int_B (r^2\ln r)_{pq} g(q)\mathrm{d}A_q$$
$$= \int_{\partial B}\left[\frac{\partial}{\partial n_Q}\{\nabla^2(r^2\ln r)_{pQ}\dot{w}(Q)\} - \nabla^2(r^2\ln r)_{pQ}\frac{\partial\dot{w}(Q)}{\partial n_Q}\right.$$
$$\left. + \frac{\partial}{\partial n_Q}(r^2\ln r)_{pQ}\nabla^2\dot{w}(Q) - (\mathrm{r}^2\ln r)_{pQ}\frac{\partial}{\partial n_Q}\{\nabla^2\dot{w}(Q)\}\right]\mathrm{d}c_Q \qquad (8.15)$$

This equation contains four quantities, $\dot{w}$, $\partial\dot{w}/\partial n$, $\nabla^2\dot{w}$ and $\partial/\partial n(\nabla^2\dot{w})$ on the boundary. Two equations relating these quantities on the boundary are obtained from the boundary conditions (e.g. eqn. (8.8) or (8.9)).

Thus, one more equation is needed. This is obtained by observing that $\psi = \nabla^2 \dot{w}$ satisfies the Poisson equation $\nabla^2 \psi = g$. Thus, proceeding exactly as before for the torsion problem, an equation analogous to (7.11) can be immediately obtained as

$$2\pi\nabla^2\dot{w}(p) - \int_B \ln r_{pq} g(q)\mathrm{d}A_q = \int_{\partial B}\left[\frac{\partial \ln r_{pQ}}{\partial n_Q}\nabla^2 \dot{w}(Q) - \ln r_{pQ}\frac{\partial}{\partial n_Q}\{\nabla^2\dot{w}(Q)\}\right]\mathrm{d}c_Q \qquad (8.16)$$

At a point P on the boundary where it is locally smooth, the boundary integral equations corresponding to (8.15) and (8.16) become

$$4\pi\dot{w}(P) - \int_B (r^2 \ln r)_{Pq} g(q)\mathrm{d}A_q = \int_{\partial B}\left[\frac{\partial}{\partial n_Q}\left\{\nabla^2(r^2\ln r)_{PQ}\dot{w}(Q) - \nabla^2(r^2\ln r)_{PQ}\frac{\partial \dot{w}(Q)}{\partial n_Q}\right.\right.$$
$$\left.\left. + \frac{\partial}{\partial n_Q}(r^2\ln r)_{PQ}\nabla^2\dot{w}(Q) - (r^2\ln r)_{PQ}\frac{\partial}{\partial n_Q}(\nabla^2\dot{w}(Q)\right\}\right]\mathrm{d}c_Q \qquad (8.17)$$

$\pi\nabla^2\dot{w}(P)$

$$\pi\dot{w}(P) - \int_B \ln r_{Pq} g(q)\mathrm{d}A_q = \int_{\partial B}\left[\frac{\partial}{\partial n_Q}(\ln r_{PQ})\nabla^2\dot{w}(Q) - \ln r_{PQ}\frac{\partial}{\partial n_Q}\{\nabla^2\dot{w}(Q)\}\right]\mathrm{d}c_Q \qquad (8.18)$$

Equations (8.17) and (8.18) are a pair of coupled integral equations relating $\dot{w}$, $\partial\dot{w}/\partial n$, $\nabla^2\dot{w}$ and $\partial/\partial n(\nabla^2\dot{w})$ on the boundary. As mentioned earlier, the other two equations must be obtained from the boundary conditions.

8.2.2 An Indirect Formulation

An indirect boundary element formulation can be obtained by using a pair of fundamental solutions of the biharmonic equation (8.6). Some simple fundamental solutions are $r^2\ln r$, $\partial/\partial n(r^2\ln r)$, $\ln r$ and $\partial\ln r/\partial n$, where the normal derivatives are taken with respect to the outward normal at the field point. Using the first and third solution, for example, results in an integral representation of the solution of eqn. (8.6) as

$$8\pi\dot{w}(p) = \int_{\partial B}(r^2\ln r)_{pQ} C_1(Q)\mathrm{d}c_Q$$

$$+\int_{\partial B}(\ln r)_{pQ}C_2(Q)\mathrm{d}c_Q$$

$$+\int_{B}(r^2\ln r)_{pq}g(q)\mathrm{d}A_q \qquad (8.19)$$

where C_1 and C_2 are distributions of boundary concentrations that must be determined by using the boundary conditions. The boundary integral equations to be used depend on the specified boundary conditions. Thus, if, for example, $\dot{w}$ and $\partial\dot{w}/\partial n$ are prescribed on ∂B, the appropriate boundary integral equations for C_1 and C_2 are

$$8\pi\dot{w}(P)=\int_{\partial B}(r^2\ln r)_{PQ}C_1(Q)\mathrm{d}c_Q$$

$$+\int_{\partial B}(\ln r)_{PQ}C_2(Q)\mathrm{d}c_Q$$

$$+\int_{B}(r^2\ln r)_{Pq}g(q)\mathrm{d}A_q \qquad (8.20)$$

$$8\pi\frac{\partial\dot{w}(P)}{\partial n_P}=\int_{\partial B}\frac{\partial}{\partial n_P}(r^2\ln r)_{PQ}C_1(Q)\mathrm{d}c_Q$$

$$+\int_{\partial B}\frac{\partial}{\partial n_P}(\ln r)_{PQ}C_2(Q)\mathrm{d}c_Q$$

$$-\pi C_2(P)+\int_{B}\frac{\partial}{\partial n_P}(r^2\ln r)_{Pq}g(q)\mathrm{d}A_q \qquad (8.21)$$

The residue term in eqn. (8.21) is negative. This is consistent with eqn. (7.16) where the tangential stress is the negative normal derivative of the stress function and $H_{3j}t_j=-\partial\ln r/\partial n$.

A comment is made here regarding the choice of fundamental solutions of the biharmonic equation. While any two of these solutions can be used, it may be advisable, generally, to use $r^2\ln r$ and $(\partial/\partial n_Q)(r^2\ln r)$, especially if higher order derivatives of the solution function are required in the analysis. This stems from the fact that both these functions are regular as $r\to 0$ so that their derivatives, although they might become singular, are better behaved than, for example, the derivatives of $\ln r$ or $\partial/\partial n_Q(\ln r)$. This has been the experience of the author when trying to solve planar fracture problems described in chapter 10. In this chapter, eqns. (8.19)–(8.21) are used to solve an illustrative analytical example for the bending of a circular elastic plate and the direct formulation is used to get numerical results for the bending of viscoplastic plates.

8.3 ANALYTICAL EXAMPLE FOR THE BENDING OF A CIRCULAR ELASTIC PLATE

The analytical example considered here is that of transverse deflection of a thin, elastic, circular plate of radius a and uniform thickness h. The plate is clamped on its boundary and is subjected to a uniform pressure q per unit area of the plate.

The geometry of the problem is the same as that shown in Fig. 5.2 (without, of course, the concentrated load in the radial direction at p). The boundary conditions are

$$w=\frac{\partial w}{\partial n}=0 \text{ at } R=a \tag{8.22}$$

The indirect BEM formulation of the previous section is used here. The relevant equations are (8.19)–(8.21) for the deflection w (rather than its rate) and, in the absence of nonelastic deformation, $g=q/D$. Thus, eqn. (8.19) for the displacement at an internal point p is

$$\begin{aligned}8\pi w(R)=&C_1(a)\int_0^{2\pi} K_1(R;a,\theta)a\mathrm{d}\theta\\&+C_2(a)\int_0^{2\pi} K_2(R;a,\theta)a\mathrm{d}\theta\\&+\int_0^a\int_0^{2\pi} K_1(R;\rho,\theta)\frac{q}{D}\rho\mathrm{d}\theta\mathrm{d}\rho\end{aligned} \tag{8.23}$$

where the kernels are defined as

$$K_1(R;\rho,\theta)=r^2\ln r$$
$$K_2(R;\rho,\theta)=\ln r$$

and

$$r=(\rho^2+R^2-2\rho R\cos\theta)^{1/2}$$

At a point P on the boundary $(R=a)$ the equations for w and $\partial w/\partial n$ become

$$\begin{aligned}w(a)=0=&C_1(a)\int_0^{2\pi} K_1(a;a,\theta)a\mathrm{d}\theta\\&+C_2(a)\int_0^{2\pi} K_2(a;a,\theta)a\mathrm{d}\theta\\&+\int_0^a\int_0^{2\pi} K_1(a;\rho,\theta)\frac{q}{D}\rho\mathrm{d}\theta\mathrm{d}\rho\end{aligned} \tag{8.24}$$

$$\frac{dw}{dR}(a)=0=C_1(a)\int_0^{2\pi}\frac{\partial K_1}{\partial R}(a;a,\theta)a d\theta$$
$$+C_2(a)\int_0^{2\pi}\frac{\partial K_2}{\partial R}(a;a,\theta)a d\theta$$
$$+\int_0^a\int_0^{2\pi}\frac{\partial K_1}{\partial R}(a;\rho,\theta)\frac{q}{D}\rho d\theta d\rho$$
$$-\pi C_2(a) \tag{8.25}$$

where
$$\frac{\partial K_1}{\partial R}(R;\rho,\theta)=(R-\rho\cos\theta)(1+2\,ln\,r)$$

and
$$\frac{\partial K_2}{\partial R}(R;\rho,\theta)=\frac{R-\rho\cos\theta}{r^2}$$

The integrals of the kernels for various cases are shown in Table 8.1. Using these, two simultaneous equations are obtained for $C_1(a)$ and $C_2(a)$.

Solving these equations,

$$C_1(a)=-\frac{qa(4\ln a+3)}{8D(1+\ln a)} \tag{8.26}$$

$$C_2(a)=\frac{a^3q\{1+2\ln a+2(\ln a)^2\}}{8D(1+\ln a)\ln a} \tag{8.27}$$

Equation (8.23) for $w(R)$ at an internal point becomes

$$8\pi w(R)=2\pi aC_1(a)\{(a^2+R^2)\ln a+R^2\}$$
$$+2\pi a\ln a C_2(a)$$
$$+\frac{\pi q}{2D}(3R^4\ln R+R^4)$$
$$+\frac{2\pi q}{D}\left[\frac{a^4\ln a}{4}+\frac{a^2R^2\ln a}{2}-\frac{3R^4\ln R}{4}\right.$$
$$\left.-\frac{3R^4}{16}+\frac{a^2R^2}{4}-\frac{a^4}{16}\right] \tag{8.28}$$

Substituting for $C_1(a)$ and $C_2(a)$ from eqns. (8.26) and (8.27) into eqn. (8.28) gives

$$w(R)=\frac{q(a^2-R^2)^2}{64D} \tag{8.29}$$

which is the well known expression from reference 1.

TABLE 8.1

INTEGRALS FOR THE BENDING OF A CIRCULAR PLATE

	$\int_0^{2\pi} f(\theta)\mathrm{d}\theta$		
$f(\theta)$	$\rho > R$	$\rho = R$	$\rho < R$
$K_1 = r^2 \ln r$	$2\pi\{R^2 + (\rho^2 + R^2)\ln\rho\}$	$2\pi R^2(1 + 2\ln R)$	$2\pi\{\rho^2 + (\rho^2 + R^2)\ln R\}$
$K_2 = \ln r$	$2\pi \ln\rho$	$2\pi \ln R$	$2\pi \ln R$
$\frac{\partial K_1}{\partial R} = (R - \rho\cos\theta)(1 + 2\ln r)$	$4\pi R(1 + \ln\rho)$	$4\pi R(1 + \ln R)$	$2\pi(R + 2R\ln R + \rho^2/R)$
$\frac{\partial K_2}{\partial R} = \frac{R - \rho\cos\theta}{r^2}$	0	$\frac{\pi}{R}$	$\frac{2\pi}{R}$

$r^2 = \rho^2 + R^2 - 2\rho R\cos\theta$

8.4 NUMERICAL IMPLEMENTATION AND SOLUTION STRATEGY

A numerical implementation and solution strategy using the direct formulation is given next. This strategy has been used to obtain the numerical results presented in the next section.

8.4.1 Discretized Equations

A discretized version of eqn. (8.18) is obtained by dividing the boundary ∂B of the plate into N_s straight boundary elements using $N_b (N_b = N_s)$ boundary nodes and the interior B of the plate into n_i triangular internal elements

$$\pi\nabla^2\dot{w} - \sum_{n_i}\int_{\Delta A_i}(\ln r)g\mathrm{d}A = \sum_{N_s}\int_{\Delta c_i}\left\{\frac{\partial}{\partial n}(\ln r)\nabla^2\dot{w} - (\ln r)\frac{\partial}{\partial n}(\nabla^2\dot{w})\right\}\mathrm{d}c \quad (8.30)$$

A similar equation results from eqn. (8.17).

The quantities $\dot{w}$, $\partial\dot{w}/\partial n$, $\nabla^2\dot{w}$ and $\partial/\partial n(\nabla^2\dot{w})$ are assumed to vary linearly over each boundary element with their values assigned (generally) at the nodes which lie at the intersections of the elements. For boundary elements adjacent to a corner, the source (sampling) points are placed slightly away from the corner itself, in order to avoid taking a limit at a point on the boundary where it is not locally smooth. This method also takes care of possible discontinuities in normal derivatives across a corner.

Integrals through the thickness of the type

$$I(x_1, x_2, t) = \int_{-h/2}^{h/2}(\dot{\varepsilon}_{11}^{(n)} + \nu\dot{\varepsilon}_{22}^{(n)})x_3\mathrm{d}x_3$$

are evaluated by Gaussian quadrature. Six Gauss points are used in the numerical calculations presented later. As mentioned earlier, the integrands are known at any time through the constitutive equations. A quadratic interpolation of such integrals is used on each triangular internal cell in the x_1x_2 plane. This FEM type methodology is more accurate than finite difference techniques for the determination of second partial derivatives of such integrals that occur in eqn. (8.7). However, the topology of the internal elements is simpler than for FEM in that every node need not be shared by neighboring elements.

Substitution of the functional forms of the variables into eqn. (8.30) and the discretized version of eqn. (8.17) leads to an algebraic system of the type

$$[A]\{\dot{x}\}=[B]\{\dot{y}\}+\{\dot{d}\} \tag{8.31}$$

The matrices $[A]$ and $[B]$ depend only on the geometry, i.e. the number of boundary nodes and their distribution. They involve integrals of the type

$$\int_{\Delta c_i} (\ln r)\mathrm{d}c, \quad \int_{\Delta c_i} (\ln r)c\mathrm{d}c$$

where c is the distance measured along a boundary element. These integrals, some of which become singular when the source point coincides with the field point, are evaluated in *closed form*. They are available in Section 8.6.1 and those for the curvature rates are given in Section 8.6.2. The vector $\{\dot{x}\}$ contains unknown and the vector $\{\dot{y}\}$ known quantities on the boundary. Finally, the vector $\{\dot{d}\}$ contains integrals of the kernels $r^2 \ln r$ and $\ln r$ over the triangular internal elements, and the nonelastic strain rates $\dot{\varepsilon}_{11}^{(n)}$, $\dot{\varepsilon}_{22}^{(n)}$ and $\dot{\varepsilon}_{12}^{(n)}$ through the function g. This vector is known at any time through the constitutive relations.

8.4.2 Solution Strategy

The solution strategy is quite similar to that discussed earlier for planar and torsion problems. The initial distribution of displacements, stresses and moments is obtained by solving the corresponding elastic BEM equations. The initial rates of the nonelastic strains are obtained from the appropriate constitutive equations. Hart's model is used in the numerical examples which follow. These nonelastic strain rates, together with the prescribed boundary conditions, are used in eqn. (8.31) which is solved for the unknown components of $\dot{w}$, $\partial\dot{w}/\partial n$, $\nabla^2\dot{w}$ and $\partial/\partial n\ (\nabla^2\dot{w})$ on the boundary ∂B. The displacement rate $\dot{w}$ and its Laplacian $\nabla^2\dot{w}$ are now obtained throughout the plate from discretized versions of eqns. (8.15) and (8.16) for an internal point.

The curvature rates are next obtained pointwise analytically from the displacement rates. The quantities

$$\left(\frac{\partial^2 \dot{w}}{\partial x_1^2}-\frac{\partial^2 \dot{w}}{\partial x_2^2}\right) \text{and} \ \frac{\partial^2 \dot{w}}{\partial x_1 \partial x_2}$$

are evaluated at an internal point by analytical differentiation of eqn. (8.15) under the integral sign. The curvature rates are obtained by combining these with $\nabla^2\dot{w}$ obtained from eqn. (8.16). Further details of this procedure are given in Section 8.6.2. This method works well except for points very near or on the boundary. The method fails for source points near the boundary because the differentiated kernels are 'strongly nearly singular' in this region and integration of these kernels becomes inaccurate. This problem has been discussed earlier by Cruse[7] in connection with elasticity problems. This difficulty is overcome by obtaining the boundary curvature rates directly from the boundary rates and by interpolation in this region close to the boundary. This method has been successfully used for clamped and simply supported straight boundaries and is described in the next section.

The stress and moment rates are next obtained from eqns. (8.3) and (8.4). The time-histories of the relevant variables are now determined by stepwise integration in time. Once again, the Euler type strategy with time-step control, discussed earlier in Section 4.2.1, is used to obtain the numerical results presented later in this chapter.

8.4.3 Boundary Curvature Rates

The x_1 and x_2 derivatives, in terms of the tangential and normal derivatives, are

$$\frac{\partial}{\partial x_1} = \sin\gamma \frac{\partial}{\partial n} + \cos\gamma \frac{\partial}{\partial c}$$

$$\frac{\partial}{\partial x_2} = -\cos\gamma \frac{\partial}{\partial n} + \sin\gamma \frac{\partial}{\partial c}$$

where γ is the angle between the x_1 axis and the anticlockwise tangential direction to the boundary element. Hence,

$$\frac{\partial^2 \dot{w}}{\partial x_1^2} = \sin^2\gamma \frac{\partial^2 \dot{w}}{\partial n^2} + 2\sin\gamma\cos\gamma \frac{\partial^2 \dot{w}}{\partial n \partial c} + \cos^2\gamma \frac{\partial^2 \dot{w}}{\partial c^2} \tag{8.32}$$

$$\frac{\partial^2 \dot{w}}{\partial x_2^2} = \cos^2\gamma \frac{\partial^2 \dot{w}}{\partial n^2} - 2\sin\gamma\cos\gamma \frac{\partial^2 \dot{w}}{\partial n \partial c} + \sin^2\gamma \frac{\partial^2 \dot{w}}{\partial c^2} \tag{8.33}$$

$$\frac{\partial^2 \dot{w}}{\partial x_1 \partial x_2} = \sin\gamma\cos\gamma\left(\frac{\partial^2 \dot{w}}{\partial c^2} - \frac{\partial^2 \dot{w}}{\partial n^2}\right) + (\sin^2\gamma - \cos^2\gamma)\frac{\partial^2 \dot{w}}{\partial n \partial c} \tag{8.34}$$

For a straight clamped edge, from eqn. (8.8), $\dot{w} = \partial\dot{w}/\partial n = 0$.

Now, $\dot{w}=0\rightarrow\frac{\partial \dot{w}}{\partial c}=0$ and $\frac{\partial^2 \dot{w}}{\partial c^2}=0\rightarrow\frac{\partial^2 \dot{w}}{\partial n^2}=\nabla^2 \dot{w}$

Also, $\frac{\partial \dot{w}}{\partial n}=0\rightarrow\frac{\partial^2 \dot{w}}{\partial n\,\partial c}=0$

Therefore,

$$\frac{\partial^2 \dot{w}}{\partial x_1^2}=\nabla^2 \dot{w}\sin^2\gamma \tag{8.35}$$

$$\frac{\partial^2 \dot{w}}{\partial x_2^2}=\nabla^2 \dot{w}\cos^2\gamma \tag{8.36}$$

$$\frac{\partial^2 \dot{w}}{\partial x_1\,\partial x_2}=-(\nabla^2 \dot{w})\sin\gamma\cos\gamma \tag{8.37}$$

For a straight simply supported edge, from eqns. (8.9)–(8.10), $\dot{w}=\nabla^2 \dot{w}=0$.

Now, $$\dot{w}=0\rightarrow\frac{\partial^2 \dot{w}}{\partial c^2}=0$$

$$\nabla^2 \dot{w}=0\rightarrow\frac{\partial^2 \dot{w}}{\partial n^2}=0$$

Therefore $$\frac{\partial^2 \dot{w}}{\partial x_1^2}=2\frac{\partial}{\partial c}\left(\frac{\partial w}{\partial n}\right)\sin\gamma\cos\gamma \tag{8.38}$$

$$\frac{\partial^2 \dot{w}}{\partial x_2^2}=-2\frac{\partial}{\partial c}\left(\frac{\partial \dot{w}}{\partial n}\right)\sin\gamma\cos\gamma \tag{8.39}$$

$$\frac{\partial^2 \dot{w}}{\partial x_1\,\partial x_2}=-\frac{\partial}{\partial c}\left(\frac{\partial \dot{w}}{\partial n}\right)\cos 2\gamma \tag{8.40}$$

The Laplacian of $\dot{w}$ on the boundary is obtained directly for the clamped case from eqn. (8.31) and the curvature rates are obtained easily from eqns. (8.35)–(8.37).

For the simply supported case, $\partial\dot{w}/\partial n$ is obtained directly on the boundary and $\partial/\partial c(\partial\dot{w}/\partial n)$ is obtained by numerical differentiation along the boundary element. Equations (8.38)–(8.40) then deliver the boundary curvature rates for this case.

8.5 NUMERICAL RESULTS AND DISCUSSION

Numerical results for several sample plate bending problems for 304 stainless steel plates at 400°C, using Hart's viscoplastic constitutive model, are presented in this section. These are taken from a paper by Morjaria and the author of this book.[6]

The values of the material parameters, used in these calculations, are the same as those used before (Section 4.3) with two differences: $\mathcal{M}=0{\cdot}113\times10^9$ psi and $\dot{\varepsilon}_0=1{\cdot}553\times10^{-4}\,\text{sec}^{-1}$. The initial values of the state variables σ^* and $\varepsilon_{ij}^{(a)}$ in Hart's model are the same as in Section 4.3.

8.5.1 Results for Square Plates

Numerical results for the bending of square plates are shown in Figs. 8.1–8.4. The plate is of side 'a' = 2 inches and the plate thickness is 2% of 'a' for the clamped plate (Figs. 8.1–8.3) and 1% of 'a' for the simply supported plate in Fig. 8.4. The loading is a uniform pressure q increasing at a constant rate of 0·5 psi/sec. The boundary and internal mesh are shown as insets in the figures.

The pressure as a function of central displacement is shown in Fig. 8.1. The short time BEM solution is within 1% of the elastic solution from reference 1.

In order better to understand the behavior of the inelastic curve in Fig. 8.1, it is instructive to consider an analogous plate made of a linear viscoelastic material which is modelled as elastic in dilatation and as a three-parameter viscoelastic solid in tension. If the free spring has modulus E and the one parallel to the dashpot has a modulus $\mathcal{M}$, the long time Young's modulus and Poisson's ratio for this model are[8]

$$E_\infty=\frac{\mathcal{M}E}{\mathcal{M}+E},\quad \nu_\infty=\frac{1}{2}-\frac{E_\infty}{6K}$$

where K is the bulk modulus of the material. In this case, for the plate bending problem, the initial and final slopes, q/w_A, are proportional to the quantities $E/(1-\nu^2)$ and $E_\infty/(1-\nu_\infty^2)$ respectively.

Returning to Hart's model (Fig. 2.1), it is observed that increase of stress from zero causes deformation of the anelastic spring but substantial permanent deformation of the 'plastic' element only begins when the stress in the upper branch, $\sigma^{(a)}$, approaches the hardness σ^*. Thus, with the plastic element essentially inactive, the behavior in tension is close to that of a nonlinear three-parameter viscoelastic solid. The zero time slope of the inelastic curve in Fig. 8.1, therefore, should be pro-

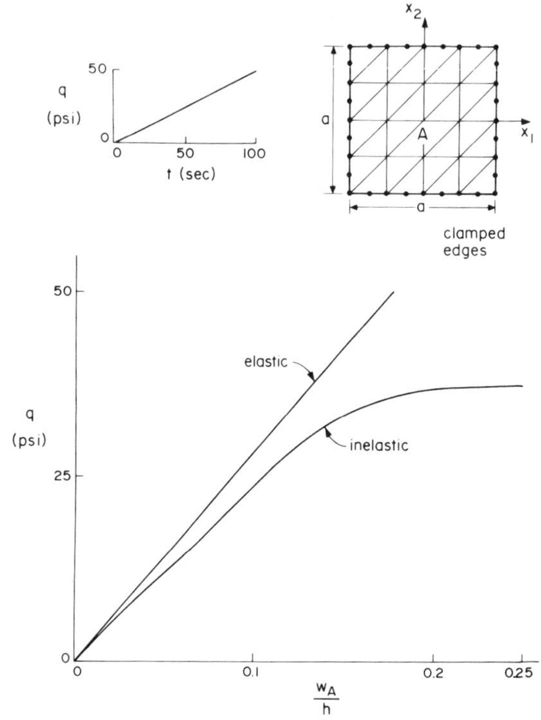

FIG. 8.1. Load as a function of central displacement for a clamped square plate subjected to a uniform load increasing at a constant rate.

portional to $E/(1-\nu^2)$ and this should change over to $E_\infty/(1-\nu_\infty^2)$ at, say, $w_A/h=0{\cdot}08$. Comparison of the numbers shows that the zero time slope is within 1% and the slope at $w_A/h=0{\cdot}08$ is within 3% of the above mentioned quantities, thus confirming the accuracy of the BEM model. Further increase of pressure causes substantial permanent deformation of the plastic element and the curve has the features of a usual elastic–plastic curve. Yielding starts at the midpoints of the outside edges of the plate. The redistribution of stress through the thickness at the midpoint of an edge of the plate (Fig. 8.2) shows the expected transition from the linear elastic distribution to the nonlinear distribution characteristic of elastic–plastic deformation. Redistribution of the nondimensionalized bending moment M_{11} along a center line of the plate ($x_2=0$) is shown in Fig. 8.3.

The load–deflection curve for a simply supported square plate is shown

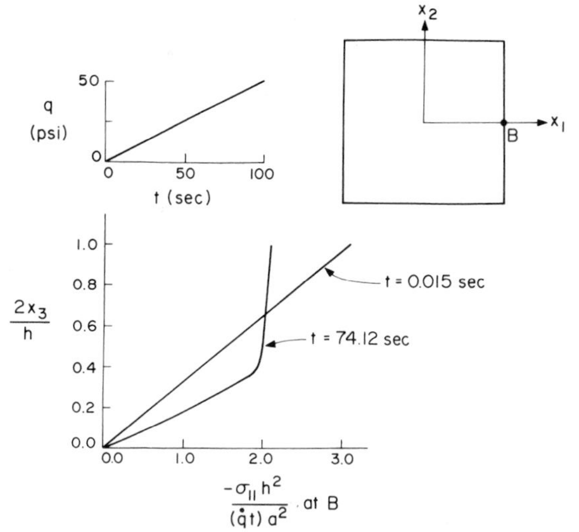

FIG. 8.2. Redistribution of stress at the mid-point of an edge of the clamped square plate of Fig. 8.1.

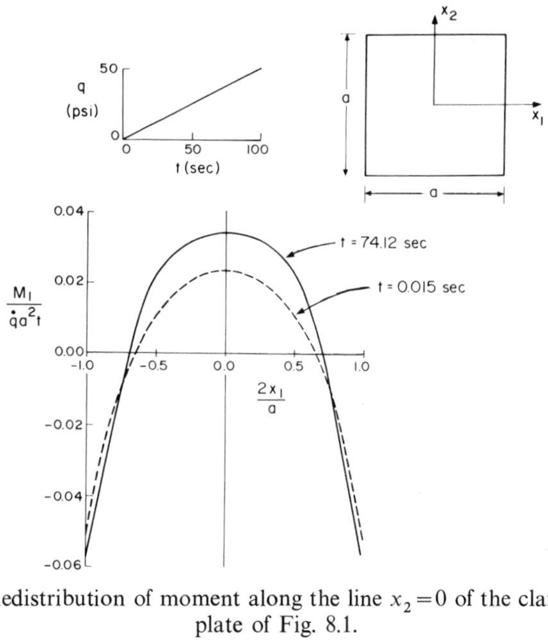

FIG. 8.3. Redistribution of moment along the line $x_2 = 0$ of the clamped square plate of Fig. 8.1.

in Fig. 8.4. This time the elastic solution is within 1% of the correct one and the slope at $w_A/h=0{\cdot}4$ is within 2% of the slope expected from the analogous viscoelastic case.

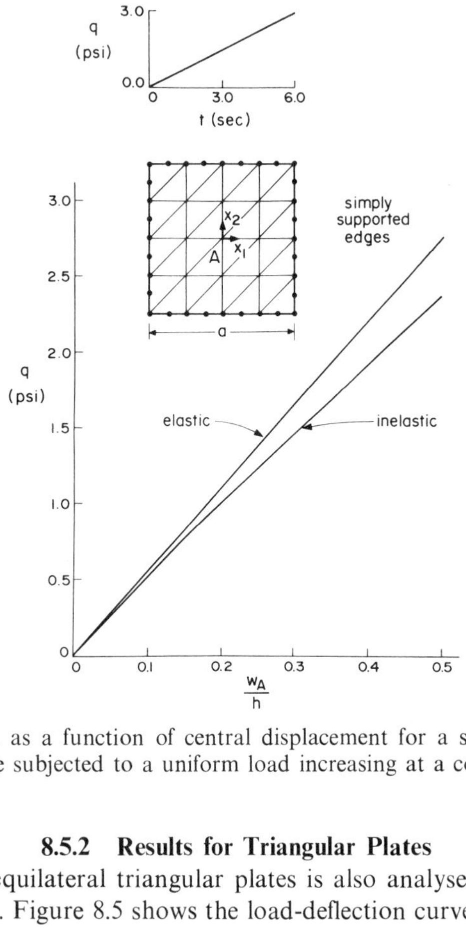

FIG. 8.4. Load as a function of central displacement for a simply supported square plate subjected to a uniform load increasing at a constant rate.

8.5.2 Results for Triangular Plates

Deflection of equilateral triangular plates is also analysed by this computer program. Figure 8.5 shows the load-deflection curve for a clamped equilateral triangular plate of side 1 inch with $h/a=2\%$. The elastic solution is within 2% of that obtained by Conway[9] and the inelastic solution has the expected elastic–plastic characteristics. Stress redistribution through the thickness (Fig. 8.6) exhibits the transition from linear elastic to nonlinear elastic–plastic analogous to Fig. 8.2. Redistribution of moment along the line $x_2=0$ is shown in Fig. 8.7. The elastic solution

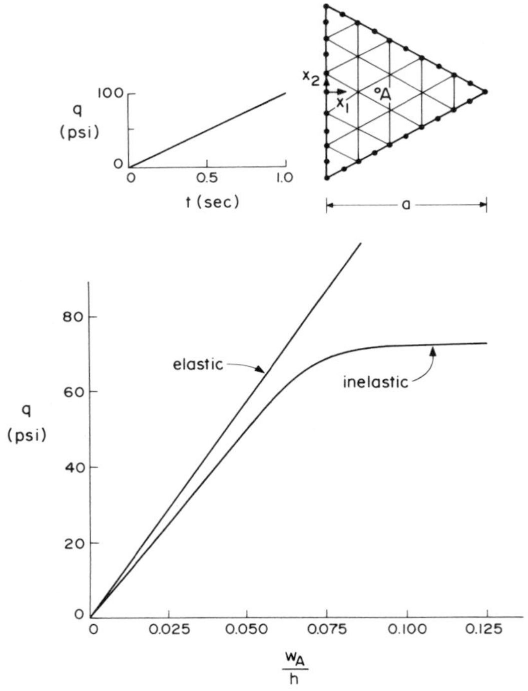

FIG. 8.5. Load as a function of centroidal displacement for a clamped triangular plate subjected to a uniform load increasing at a constant rate.

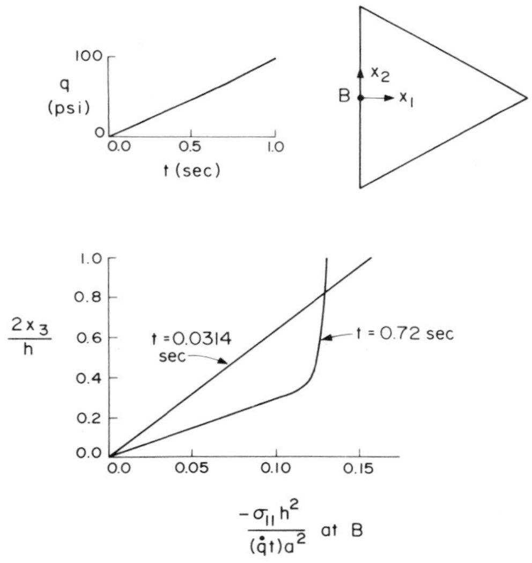

FIG. 8.6. Redistribution of stress at the mid-point of an edge of the clamped triangular plate of Fig. 8.5.

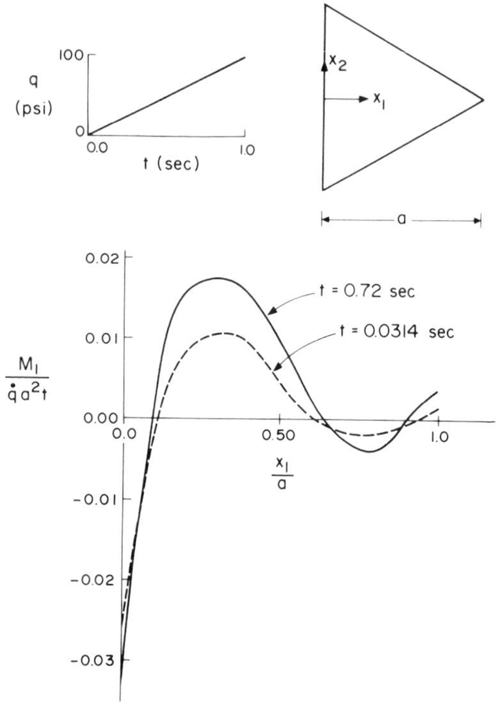

FIG. 8.7. Redistribution of moment along the line $x_2=0$ of the clamped triangular plate of Fig. 8.5.

for a simply supported equilateral triangular plate under uniform pressure, obtained by the BEM program, is within 0·04% of the elastic solution.[1] The nonelastic solution has not been determined in this case.

The number of time-steps and computer times on an IBM 370/168, required for these problems, are given in Table 8.2.

8.6 APPENDIX

8.6.1 Integrals in Equations (8.15)–(8.18)

The integrals in eqns. (8.15)–(8.18) are evaluated using a method similar to that described by Riccardella.[10] The notation used for the evaluation of boundary integrals is given in Fig. 5.12 and that for the area integrals in Figs. 5.13 and 8.8.

TABLE 8.2
BEM PROGRAM STATISTICS FOR PLATE BENDING PROBLEMS

Problem	*Number of boundary nodes*	*Number of internal elements*	*Number of time steps*	*CPU time on* IBM 370/168 (*sec*)
1. Clamped square plate	36	32	372	208
2. Simply supported square plate	36	32	350	210
3. Clamped triangular plate	33	25	200	98

The quantities $\dot{w}$, $\partial\dot{w}/\partial n$, $\nabla^2\dot{w}$ and $\partial/\partial n\,\nabla^2\dot{w}$ are interpolated linearly on each boundary segment. Thus,

$$\dot{w}(P)=\frac{\dot{w}(b)-\dot{w}(a)}{\Delta c}c+\dot{w}(a)$$

and similarly for the other quantities, where $c=D\tan\theta-z_a$ and $\dot{w}(b)$, $\dot{w}(a)$ are the displacement rates at nodes b and a respectively.

A typical integral of the following type would become

$$\int_{\Delta c}\frac{\partial}{\partial n}\{\nabla^2(r^2\ln r)\}\dot{w}\,\mathrm{d}c=\frac{\dot{w}(b)-\dot{w}(a)}{\Delta c}\int\frac{\partial}{\partial n}\{\nabla^2(r^2\ln r)\}D\tan\theta\,\mathrm{d}c$$
$$+\left[\dot{w}(a)-\frac{\{\dot{w}(b)-\dot{w}(a)\}^2a}{\Delta c}\right]\int\frac{\partial}{\partial n}\{\nabla^2(r^2\ln r)\}\mathrm{d}c$$

with $r=D/\cos\theta$

For $D\neq 0$ the integrals over a boundary segment are as follows:

1. $$A_1=\int\ln r\,\mathrm{d}c=D\{\tan\theta_b(\ln r_b-1)-\tan\theta_a(\ln r_a-1)+\theta_b-\theta_a\}$$

$$A_2=\int\ln r\,\mathrm{d}c\,D\tan\theta=\frac{r_b^2}{4}(2\ln r_b-1)-\frac{r_a^2}{4}(2\ln r_a-1)$$

2. $$B_1=\int\frac{\partial}{\partial n}(\ln r)\mathrm{d}c=\theta_b-\theta_a$$

$$B_2=\int\frac{\partial}{\partial n}(\ln r)\mathrm{d}c\,D\tan\theta=D\ln(r_b/r_a)$$

3. $$C_1=\int r^2\ln r\,\mathrm{d}c=D/3\,\{r_b^2\tan\theta_b\ln r_b-r_a^2\tan\theta_a\ln r_a$$
$$+D^2(2\tan\theta_b\ln r_b-2\tan\theta_a\ln r_a)$$
$$-\frac{D^2}{3}(\tan^3\theta_b-\tan^3\theta_a)-2D^2(\tan\theta_b-\tan\theta_a)$$
$$+2D^2(\theta_b-\theta_a)\}$$

$$C_2=\int r^2\ln r\,\mathrm{d}c\;D\tan\theta=\frac{r_b^4}{4}\ln r_b-\frac{r_a^4}{4}\ln r_a-\frac{1}{16}(r_b^4-r_a^4)$$

4. $$D_1=\int\frac{\partial}{\partial n}(r^2\ln r)\mathrm{d}c=D(2A_1+\Delta s)$$

$$D_2=\int\frac{\partial}{\partial n}(r^2\ln r)\mathrm{d}c\,D\tan\theta=D\left\{2A_2+\tfrac{1}{2}(r_b^2-r_a^2)\right\}$$

5. $$E_1 = \int \nabla^2 (r^2 \ln r) \mathrm{d}c = 4(A_1 + \Delta s)$$

$$E_2 = \int \nabla^2 (r^2 \ln r) \mathrm{d}c\, D \tan\theta = 4\{A_2 + \tfrac{1}{2}(r_b^2 - r_a^2)\}$$

6. $$F_1 = \int \frac{\partial}{\partial n}\{\nabla^2 (r^2 \ln r)\}\mathrm{d}c = 4(\theta_b - \theta_a)$$

$$F_2 = \int \frac{\partial}{\partial n}\{\nabla^2 (r^2 \ln r)\}\mathrm{d}c\; D\tan\theta = 4D\ln(r_b/r_a)$$

For $D=0$ $\mathrm{SGN} = \sin\theta$

1. $$A_1 = \mathrm{SGN}\{r_b(\ln r_b - 1) - r_a(\ln r_a - 1)\}$$
$$A_2 = \{r_b^2(2\ln r_b - 1) - r_a^2(2\ln r_a - 1)\}/4$$

2. $$B_1 = 0$$
$$B_2 = 0$$

3. $$C_1 = \mathrm{SGN}\{r_b^3(3\ln r_b - 1) - r_a^3(3\ln r_a - 1)\}/9$$
$$C_2 = \{r_b^4(4\ln r_b - 1) - r_a^4(4\ln r_a - 1)\}/16$$

4. $$D_1 = 0$$
$$D_2 = 0$$

5. $$E_1 = 4(A_1 + \Delta c)$$
$$E_2 = 4\{A_2 + \frac{1}{2}(r_b^2 - r_a^2)\}$$

6. $$F_1 = 0$$
$$F_2 = 0$$

$$\int \frac{\partial}{\partial n}\nabla^2(r^2 \ln r)\dot{w}\mathrm{d}c = \frac{\dot{w}(b)}{\Delta c}(F_2 - z_a F_1) + \dot{w}(a)\left\{F_1\left(1 + \frac{z_a}{\Delta c}\right) - \frac{F_2}{\Delta c}\right\}$$

Area Integrals of the type

$$\int_{\Delta A_i} \ln r g \mathrm{d}A$$

are obtained by summing algebraically the contribution of each triangle formed by joining a side of the triangular element to the source point as shown in Fig. 8.8. The quantity g is assumed to be constant over the triangular element.

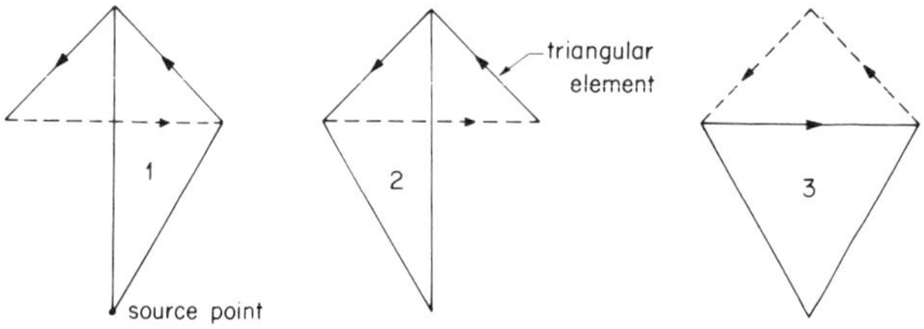

FIG. 8.8. Area integration scheme for triangular internal elements.

7. $$G=g\int_A \ln r\mathrm{d}A=\frac{g}{4}\{2DA_1-D^2(\tan\theta_b-\tan\theta_a)\}$$

8. $$H=g\int r^2\ln r\mathrm{d}A$$
$$=D[4C_1-D^3\{(\tan^3\theta_b-\tan^3\theta_a)/3+(\tan\theta_b-\tan\theta_a)\}]/16$$

8.6.2 Evaluation of Curvatures

By analytical differentiation of eqn. (8.15) the following can be obtained

$$4\pi\left(\frac{\partial^2\dot{w}}{\partial x_1^2}-\frac{\partial^2\dot{w}}{\partial x_2^2}\right)-\int_A\frac{\xi^2-\eta^2}{r^2}g\mathrm{d}A=\int_{\partial B}\left\{-4\dot{w}\frac{\partial}{\partial n}\left(\frac{\xi^2-\eta^2}{r^4}\right)+4\frac{\partial\dot{w}}{\partial n}\left(\frac{\xi^2-\eta^2}{r^4}\right)\right.$$
$$\left.+\nabla^2\dot{w}\frac{\partial}{\partial n}\left(\frac{\xi^2-\eta^2}{r^2}\right)-\frac{\partial}{\partial n}\nabla^2\dot{w}\left(\frac{\xi^2-\eta^2}{r^2}\right)\right\}\mathrm{d}c$$

$$4\pi\frac{\partial^2\dot{w}}{\partial x_1\partial x_2}-\int_A\frac{\xi\eta}{r^2}g\mathrm{d}A=\int_{\partial B}\left\{-4\dot{w}\frac{\partial}{\partial n}\frac{\xi\eta}{r^4}+4\frac{\partial\dot{w}}{\partial n}\frac{\xi\eta}{r^4}\right.$$
$$\left.+\nabla^2\dot{w}\frac{\partial}{\partial n}\frac{\xi\eta}{r^2}-\frac{\partial}{\partial n}(\nabla^2\dot{w})\frac{\xi\eta}{r^2}\right\}\mathrm{d}c$$

where ξ, η are the x_1 and x_2 distances from the source point to the field point as shown in Fig. 8.9. The relevant integrals used here are

9. $$I_1=\int\frac{\xi^2-\eta^2}{r^2}\mathrm{d}c$$
$$=D\{2\cos 2\alpha(\theta_b-\theta_a)-\cos 2\alpha(\tan\theta_b-\tan\theta_a)\}-2\sin 2\alpha\ln(r_b/r_a)$$
$$I_2=\int\frac{\xi^2-\eta^2}{r^2}\mathrm{d}c\,D\tan\theta$$
$$=D^2\left\{2\cos 2\alpha\ln\frac{r_b}{r_a}-\frac{\cos 2\alpha}{2}(\tan^2\theta_b-\tan^2\theta_a)\right.$$

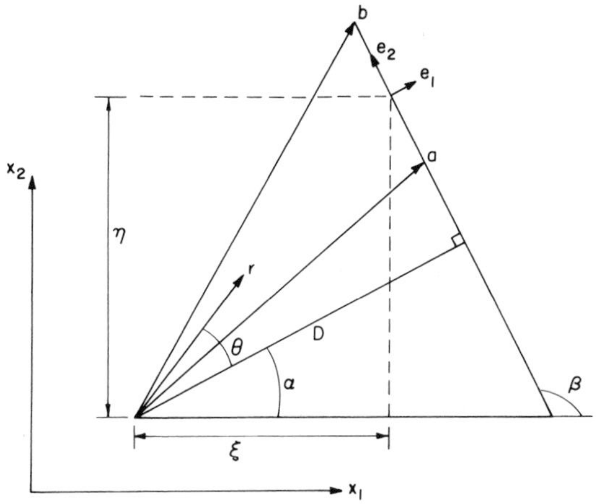

FIG. 8.9. Notation used for the evaluation of integrals for calculating curvatures.

$$-2\sin 2\alpha(\tan\theta_b-\tan\theta_a-\theta_b+\theta_a)\Big\}$$

10. $$J_1=\int\frac{\partial}{\partial n}\left(\frac{\xi^2-\eta^2}{r^2}\right)\mathrm{d}c$$

$$=2\cos 2\alpha(\theta_b-\theta_a)-2\sin 2\alpha\ln(r_b/r_a)-\sin 2(\theta_b+\alpha)+\sin 2(\theta_a+\alpha)$$

$$J_2=\int\frac{\partial}{\partial n}\left(\frac{\xi^2-\eta^2}{r^2}\right)\mathrm{d}cD\tan\theta=2D\{\cos 2\alpha[\cos 2\theta_b-\cos 2\theta_a$$

$$+4\ln(r_b/r_a)]/2$$

$$+\sin 2\alpha[2(\theta_b-\theta_a)-(\tan\theta_b-\tan\theta_a)$$

$$-\tfrac{1}{2}(\sin 2\theta_b-\sin 2\theta_a)]\}$$

11. $$K_1=\int\frac{\xi^2-\eta^2}{r^4}\mathrm{d}c$$

$$=\frac{1}{2D}\{\sin 2(\theta_b+\alpha)-\sin 2(\theta_a+\alpha)\}$$

$$K_2=\int\left(\frac{\xi^2-\eta^2}{r^4}\right)\mathrm{d}c\,D\tan\theta$$

$$=\sin 2\alpha(\sin 2\theta_b-\sin 2\theta_a)/2-\sin 2\alpha(\theta_b-\theta_a)-\cos 2\alpha\{\cos(2\theta_b)$$

$$-\cos(2\theta_a)\}/2$$

$$-\cos 2\alpha\ln(r_b/r_a)$$

12. $$L_1=\int\frac{\partial}{\partial n}\left(\frac{\xi^2-\eta^2}{r^4}\right)\mathrm{d}c$$

$$=-\frac{1}{4D^2}\{2\sin 2(\theta_b+\alpha)-2\sin 2(\theta_a+\alpha)+\sin(4\theta_b+2\alpha)$$

$$-\sin(4\theta_a+2\alpha)\}$$

$$L_2=\int\frac{\partial}{\partial n}\left(\frac{\xi^2-\eta^2}{r^4}\right)\mathrm{d}c\,D\tan\theta$$

$$=\frac{1}{4D}\{\cos(4\theta_b+2\alpha)-\cos(4\theta_a+2\alpha)-2\cos(2\theta_b+2\alpha)$$

$$+2\cos(2\theta_a+2\alpha)\}$$

13. $$M_1=\int\frac{\xi\eta}{r^2}\mathrm{d}c$$

$$=D/2[2\cos 2\alpha\ln(r_b/r_a)+\sin 2\alpha\{2(\theta_b+\theta_a)-(\tan\theta_b-\tan\theta_a)\}]$$

$$M_2=\int\frac{\xi\eta}{r^2}\mathrm{d}c\,D\tan\theta$$

$$=D^2[\cos 2\alpha(\tan\theta_b-\tan\theta_a-\theta_b-\theta_a)+\sin 2\alpha\{\ln(r_b/r_a)$$

$$-\tfrac{1}{4}(\tan^2\theta_b-\tan^2\theta_a)\}]$$

14. $$N_1=\int\frac{\partial}{\partial n}\left(\frac{\xi\eta}{r^2}\right)\mathrm{d}c$$

$$=\cos 2\alpha\ln(r_b/r_a)+\sin 2\alpha(\theta_b-\theta_a)+\tfrac{1}{2}\{\cos(2\theta_b+2\alpha)$$

$$-\cos(2\theta_a+2\alpha)\}$$

$$N_2=\int\frac{\partial}{\partial n}\left(\frac{\xi\eta}{r^2}\right)\mathrm{d}c\,D\tan\theta$$

$$=D[\cos 2\alpha\{(\tan\theta_b-\tan\theta_a-2(\theta_b-\theta_a)+\tfrac{1}{2}(\sin 2\theta_b-\sin 2\theta_a)\}$$

$$+\sin 2\alpha\{2\ln(r_b/r_a)+\tfrac{1}{2}(\cos 2\theta_b-\cos 2\theta_a)\}]$$

15. $$O_1 = \int \frac{\xi\eta}{r^4} \mathrm{d}c$$

$$= -\frac{1}{4D}\{\cos 2(\theta_b + \alpha) - \cos 2(\theta_a + \alpha)\}$$

$$O_2 = \int \frac{\xi\eta}{r^4} \mathrm{d}c\, D \tan\theta$$

$$= \tfrac{1}{2}[\cos 2\alpha\{(\theta_b - \theta_a) - \tfrac{1}{2}(\sin 2\theta_b - \sin 2\theta_a)\}$$

$$- \sin 2\alpha\{\tfrac{1}{2}(\cos 2\theta_b - \cos 2\theta_a) + \ln r_b/r_a\}]$$

16. $$P_1 = \int \frac{\partial}{\partial n}\left(\frac{\xi\eta}{r^4}\right) \mathrm{d}c$$

$$= \frac{1}{8D^2}\{\cos(4\theta_b + 2\alpha) - \cos(4\theta_a + 2\alpha)$$

$$+ 2[\cos(2\theta_b + 2\alpha) - \cos(2\theta_a + 2\alpha)]\}$$

$$P_2 = \int \frac{\partial}{\partial n}\left(\frac{\xi\eta}{r^4}\right) \mathrm{d}c\, D \tan\theta = \frac{1}{8D}\{\sin(4\theta_b + 2\alpha)$$

$$- \sin(4\theta_a + 2\alpha)$$

$$- 2[\sin(2\theta_b + 2\alpha) - \sin(2\theta_a + 2\alpha)]\}$$

8.6.3 Proof of $\nabla^2 \dot{w} = 0$ in Simple Supported Boundary

Using eqns. (8.3) and (8.4), the moment rates can be written as

$$\dot{M}_1 = D(\dot{\kappa}_1 + \nu\dot{\kappa}_2) - \frac{E}{1-\nu^2}\int_{-h/2}^{h/2} (\dot{\varepsilon}_{11}^{(n)} + \nu\dot{\varepsilon}_{22}^{(n)}) x_3 \mathrm{d}x_3$$

$$\dot{M}_2 = D(\dot{\kappa}_2 + \nu\dot{\kappa}_1) - \frac{E}{1-\nu^2}\int_{-h/2}^{h/2} (\dot{\varepsilon}_{22}^{(n)} + \nu\dot{\varepsilon}_{11}^{(n)}) x_3 \mathrm{d}x_3$$

with

$$\dot{\kappa}_1 = -\frac{\partial^2 \dot{w}}{\partial x_1^2}, \quad \dot{\kappa}_2 = -\frac{\partial^2 \dot{w}}{\partial x_2^2}$$

For the elastic problem

$$M_1 = D(\kappa_1 + \nu\kappa_2),\ M_2 = D(\kappa_2 + \nu\kappa_1)$$

$$\sigma_{11} = \frac{12}{h^3}M_1 x_3,\ \sigma_{22} = \frac{12}{h^3}M_2 x_3$$

On a simply supported straight edge parallel to the x_2 axis, for the elastic problem, the stresses and moments are zero, i.e.:

$$M_1 = 0,\ \frac{\partial^2 w}{\partial x_2^2} = 0 \rightarrow \frac{\partial^2 w}{\partial x_1^2} = 0,\ M_2 = 0,\ \sigma_{11} = 0,\ \sigma_{22} = 0$$

Now for the inelastic problem,

$$\sigma_{11} = 0,\ \sigma_{22} = 0,\ \sigma_{33} = 0 \rightarrow \dot{\varepsilon}_{11}^{(n)} = \dot{\varepsilon}_{22}^{(n)} = 0,\ \text{for all time, provided}$$

$$\varepsilon_{ij}^{(a)}(\mathbf{x}, 0) = 0$$

Now $$\dot{\varepsilon}_{11}^{(n)} = 0,\quad \dot{\varepsilon}_{22}^{(n)} = 0,\quad \dot{M}_1 = 0,\quad \frac{\partial^2 \dot{w}}{\partial x_2^2} = 0 \rightarrow \frac{\partial^2 \dot{w}}{\partial x_1^2} = 0,$$

$$\nabla^2 \dot{w} = 0$$

It can be shown that $\nabla^2 \dot{w} = 0$ for a simply supported edge of arbitrary orientation.

REFERENCES

1. TIMOSHENKO, S. P. and WOINOWSKY-KRIEGER, S. *Theory of Plates and Shells*, 2nd ed., McGraw-Hill, New York (1959).
2. JASWON, M. A. and MAITI, M. An integral equation formulation of plate bending problems. *Journal of Engineering Mathematics*, **2**, 83–93 (1968).
3. BENZINE, G. P. and GAMBY, D. A. A new integral equation formulation for plate bending problems, *Recent Advances in Boundary Element Methods*, C. A. Brebbia (ed.), Pentech Press, London and Plymouth, 327–342 (1978).
4. STERN, M. A general boundary integral formulation for the numerical solution of plate bending problems. *International Journal of Solids and Structures*, **15**, 769–782 (1979).
5. RZASNICKI, W. and MENDELSON, A. Application of the boundary integral method to elastoplastic analysis of V-notched beams. *International Journal of Fracture*, **11**, 329–342 (1975).
6. MORJARIA, M. and MUKHERJEE, S. Inelastic analysis of the transverse deflection of plates by the boundary element method. American Society of Mechanical Engineers, *Journal of Applied Mechanics*, **47**, 291–296 (1980).
7. CRUSE, T. A. Numerical solutions in three-dimensional elastostatics. *International Journal of Solids and Structures*, **5**, 1259–1274 (1969).

8. FLÜGGE, W. *Viscoelasticity*, 1st ed., Blaisdell, Waltham, Massachusetts (1967).
9. CONWAY, H. D. The approximate analysis of certain boundary-value problems. American Society of Mechanical Engineers, *Journal of Applied Mechanics*, **27**, 275–277 (1960).
10. RICCARDELLA, P. C. *An Implementation of the Boundary-Integral Technique for Planar Problems in Elasticity and Elastoplasticity*, Report No. SM-73-10, Department of Mechanical Engineering, Carnegie Mellon University, Pittsburg, PA (1973).

CHAPTER 9

Cracked Plates Loaded in Anti-plane Shear

The nonelastic deformation of cracked plates loaded in anti-plane shear is discussed in this chapter. The plate material is modelled by an elastic power law creep constitutive model. The standard boundary element method discussed earlier for planar nonelastic problems can, in principle, be used for this problem. This approach, however, can present numerical difficulties if applied to bodies with sharp cutouts or cracks. This is because the modelling of sharp stress gradients near crack-tips can require a prohibitively large number of boundary elements on the crack surface. An alternative approach using augmented kernels has proved to be very efficient. In this approach, the singular kernels for an infinite domain are augmented so that the new kernels are singular solutions for infinite domains with cutouts, with the proper boundary conditions (e.g. traction free) satisfied on the cutout surface. Thus, the effect of the cutout on the stress and displacement fields is incorporated in the kernels. This means that the unknowns of the problem lie only on the outer boundary of the body and discrete modelling of the cutout boundary is no longer necessary in a numerical solution procedure in many cases. Sometimes, as explained later in this chapter, a *known* boundary integral may have to be included on the cutout surface, but this is easy to incorporate numerically. This efficient approach is presented in this chapter. Numerical results for several sample problems are presented and discussed. Solutions for planar problems are discussed in the last chapter of this book.

9.1 GOVERNING DIFFERENTIAL EQUATIONS

A planar body, with the x_1 and x_2 axes in the plane of the body, and the x_3 axis normal to it, is loaded in anti-plane shear. The nonzero stress components are σ_{31} $(=\sigma_{13})$ and σ_{32} $(=\sigma_{23})$ and, as in the torsion problem, a stress function Φ is defined in the usual way as

$$\sigma_{31}=\frac{\partial\Phi}{\partial x_2}, \quad \sigma_{32}=-\frac{\partial\Phi}{\partial x_1} \tag{9.1}$$

The constitutive equations are the same as eqns. (7.2) and (7.3) and the compatibility equation, in this case, is

$$\frac{\partial\dot{\varepsilon}_{31}}{\partial x_2}-\frac{\partial\dot{\varepsilon}_{32}}{\partial x_1}=0 \tag{9.2}$$

Combining eqns. (7.2), (7.3), (9.1) and (9.2) results in Poisson's equation for the stress function rate in the form

$$\nabla^2\dot{\Phi}=-2G\left[\frac{\partial\dot{\varepsilon}_{31}^{(n)}}{\partial x_2}-\frac{\partial\dot{\varepsilon}_{32}^{(n)}}{\partial x_1}\right]\equiv F^{(n)} \tag{9.3}$$

Only traction boundary conditions are considered in this analysis. The traction $\tau_3=\tau$ at any point on the boundary of the body is

$$\tau=\sigma_{3i}n_i=\frac{\partial\Phi}{\partial x_2}n_1-\frac{\partial\Phi}{\partial x_1}n_2=\frac{\partial\Phi}{\partial x_i}t_i=\frac{\mathrm{d}\Phi}{\mathrm{d}c} \tag{9.4}$$

where (n_1,n_2) and (t_1,t_2) are the components of the unit outward normal and unit anticlockwise tangent to the boundary ∂B of the body at a point on it, and $\mathrm{d}\Phi/\mathrm{d}c$ is the tangential derivative of the stress function Φ. As before, c is the distance measured along the boundary in an anticlockwise sense.

9.2 BOUNDARY ELEMENT FORMULATION

9.2.1 Simply Connected Body

The Poisson equation (9.3) can be transformed, as before, to an integral equation with the kernel $\ln r$. The indirect formulation in this case gives

$$2\pi\dot{\Phi}(p)=\int_{\partial B}\ln r_{pQ}F(Q)\mathrm{d}c_Q+\int_B\ln r_{pq}F^{(n)}(q)\mathrm{d}A_q \tag{9.5}$$

where, as before in the case of torsion problems, F is the boundary density function to be determined from the boundary conditions.

This equation can alternatively be written in terms of the kernel

$$K = \mathrm{Re}\{\hat{\phi}(z, z_0)\}$$

as

$$2\pi\dot{\Phi}(p) = \int_{\partial B} K(p, Q)F(Q)\mathrm{d}c_Q + \int_B K(p, q)F^{(n)}(q)\mathrm{d}A_q \tag{9.6}$$

where

$$\hat{\phi}(z, z_0) = \ln(z - z_0),$$

Re denotes the real part of the complex argument and z and z_0 are the source and field points respectively (see Fig. 9.1).

9.2.2 Body With Cutout

A body with a cutout (Fig. 9.2) is considered next. As stated in the introduction to this chapter, problems with multiply connected regions can also be solved by the usual boundary element method. If the cutout is sharp, however, as when a crack is modelled as a line or a very thin ellipse, numerical difficulties might arise with the usual approach. In such cases, some researchers have suggested the use of an augmented kernel which includes the effect of the crack and thereby obviates the need for discrete modelling of the crack surface. Cruse[1] and Sadegh and Altiero[2,3] have used this idea for elastic problems. Mukherjee and his coworkers have applied this idea to planar and Mode III inelastic deformation problems.[4-7]

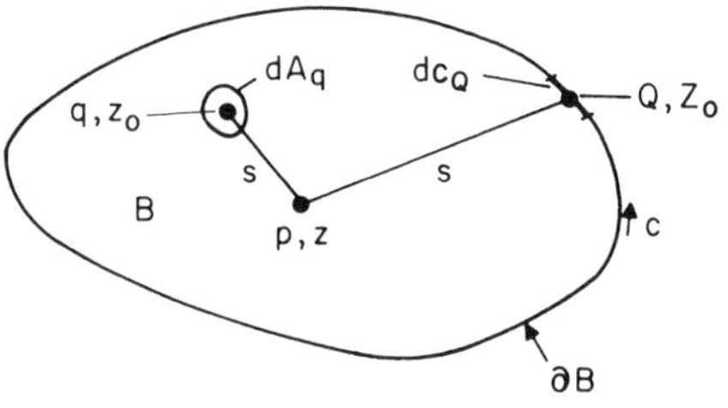

FIG. 9.1.

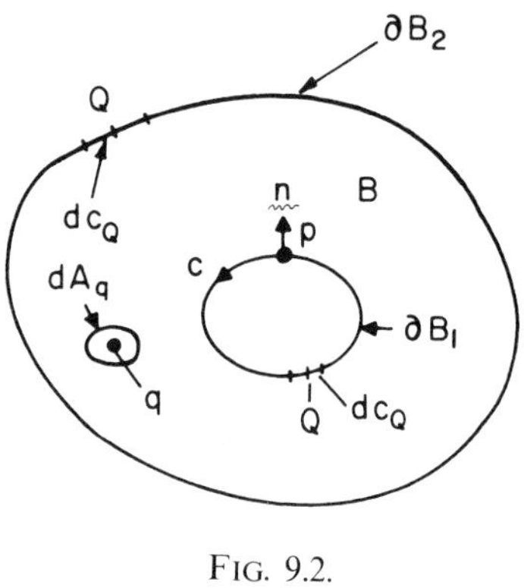

FIG. 9.2.

The augmented kernel is first derived for a body with a circular cutout. It is then extended to the case of an elliptic cutout by complex variable mapping methods.

The cutout boundary is called ∂B_1 and the outside boundary is ∂B_2, as in Fig. 9.2. If the stress function Φ is constant on ∂B_1, the traction on it is zero by virtue of eqn. (9.4). This constant can be taken to be zero without loss of generality. The cutout is first considered to be a unit circle. A function ϕ^* is defined as

$$\phi^*(\bar{z}, z_0) = -\ln\left(\frac{1}{\bar{z}} - z_0\right).$$

This function has the following properties:

(a) ϕ^* satisfies Laplace's equation.
(b) ϕ^* is regular outside the unit circle $|z|=1$.
(c) $\phi(z, \bar{z}, z_0) = \hat{\phi}(z, z_0) + \phi^*(\bar{z}, z_0)$

$$= \ln(z - z_0) - \ln\left(\frac{1}{\bar{z}} - z_0\right) \tag{9.7}$$

(d) it vanishes on the unit circle $|z|=1$.

Thus, if a kernel defined as the real part of ϕ is used in a formulation analogous to eqn. (9.6) for a body with a circular cutout, the cutout will be traction free. The derivation of ϕ^* is analogous to the circle theorem of Milne-Thomson[8] for two-dimensional irrotational incompressible inviscid flow of a fluid past a circular cylinder.

The function ϕ^* for an elliptic cutout is derived by making use of conformal mapping techniques. The mapping function

$$z = w(\xi) = m_1 \xi + \frac{m_2}{\xi} \tag{9.8}$$

transforms the region on and outside an ellipse in the z plane to the region on and inside a unit circle in the ξ plane (see Fig. 10.2). The parameters are $m_1 = (a-b)/2$, $m_2 = (a+b)/2$ in terms of the semimajor and minor axes, a and b respectively, of the ellipse. Thus, $m_1 = m_2 = a/2$ represents a line crack. Now

$$\hat{\phi} = \ln(z - z_0) = \ln(m_1 \xi + \frac{m_2}{\xi} - z_0)$$

$$= \ln m_1 + \ln(\xi - r_0) + \ln(1 - r_i/\xi) \tag{9.9}$$

where

$$r_{0,i} = \frac{z_0 \pm \sqrt{z_0^2 - 4m_1 m_2}}{2m_1}$$

are the roots of $m_1 \xi^2 - z_0 \xi + m_2 = 0$ with $|r_0| \geq 1$, $|r_i| \leq 1$ and

$$\xi = \frac{z \pm \sqrt{z^2 - 4m_1 m_2}}{2m_1} \text{ with } |\xi| \leq 1.$$

The last term in eqn. (9.9) has singularities at $\xi = 0$ and $\xi = r_i$ inside the unit circle in the ξ plane. For this case, ϕ^* is taken as

$$\phi^* = -\ln m_1 - \ln(\xi - r_0) - \ln(1 - r_i \bar{\xi})$$

which is regular inside the unit circle in the ξ plane, so that

$$\phi(z, \bar{z}, z_0) = \hat{\phi} + \phi^* = \ln(1 - r_i/\xi) - \ln(1 - r_i \bar{\xi}) \tag{9.10}$$

This function vanishes on the unit circle $|\xi| = 1$ (and therefore on the elliptical cutout in the z plane) and satisfies the other properties required of the augmented kernel.

9.2.3 Integral Equations for Stress and Traction Rates

The time-histories of the stress components σ_{31} and σ_{32} are of primary interest in this analysis. Thus, it is convenient to write integral equations directly for the stress rates using differentiated versions of the kernel ϕ. For a body with an elliptical cutout (Fig. 9.3) the equations for the rates

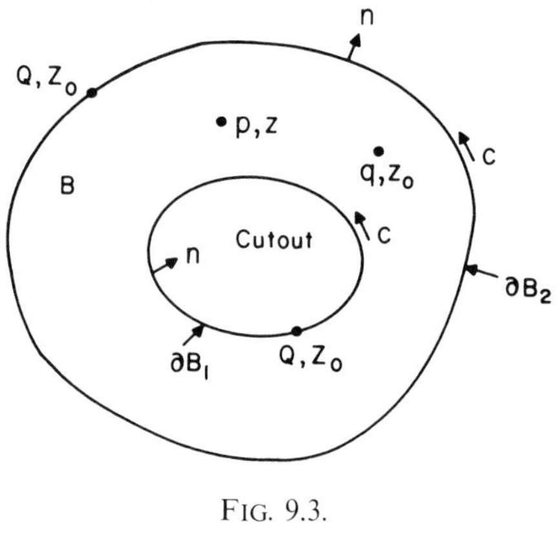

FIG. 9.3.

of stress, with $j=k=1, 2$ are

$$2\pi\dot{\sigma}_{3j}(p)=\int_{\partial B_2} H_{3j}(p,Q)F(Q)\,\mathrm{d}c_Q + \int_B H_{3j}(p,q)F^{(n)}(q)\,\mathrm{d}A_q - \int_{\partial B_1} H_{3j}(p,Q)D_K^{(n)}(Q)n_K(Q)\,\mathrm{d}c_Q \tag{9.11}$$

where $D_1^{(n)}=2G\dot{\varepsilon}_{32}^{(n)}$, $D_2^{(n)}=-2G\dot{\varepsilon}_{31}^{(n)}$, $(\nabla\cdot\mathbf{D}^{(n)}=F^{(n)})$ and the augmented kernels H_{3j} are

$$H_{31}=\mathrm{Re}\left\{\frac{\partial\phi}{\partial x_2}(z,\bar{z},z_0)\right\}=\mathrm{Im}\left(\frac{\partial\phi}{\partial\bar{z}}-\frac{\partial\phi}{\partial z}\right)$$

$$H_{32}=-\mathrm{Re}\left\{\frac{\partial\phi}{\partial x_1}(z,\bar{z},z_0)\right\}=-\mathrm{Re}\left(\frac{\partial\phi}{\partial z}+\frac{\partial\phi}{\partial\bar{z}}\right) \tag{9.12}$$

with ϕ given in eqn. (9.10).

The first two terms in eqn. (9.11) are analogous to eqn. (9.6) and the last term, which represents a layer of concentration $\mathbf{n}\cdot\mathbf{D}^{(n)}$ on the cutout boundary ∂B_1, is necessary for obtaining single-valued displacements on ∂B_1. This is proved in the next subsection.

The boundary conditions of the problem are specified in terms of the traction rates on the boundary ∂B_2. As in the torsion problem, the residue term must be included when taking the limit $p\to P$ on ∂B_2 for eqn. (9.11). This gives

$$2\pi\dot{\sigma}_{3j}(P)=\int_{\partial B_2} H_{3j}(P,Q)F(Q)\,\mathrm{d}c_Q + \int_B H_{3j}(P,q)F^{(n)}(q)\,\mathrm{d}A_q - \int_{\partial B_1} H_{3j}(P,Q)D_K^{(n)}(Q)n_K(Q)\,\mathrm{d}c_Q + \pi t_j(P)F(P) \tag{9.13}$$

where ∂B_2 is locally smooth at P.

The traction rate for a point P on ∂B_2 where it is locally smooth, is, from eqn. (9.13) (with $j, k=1, 2$)

$$2\pi\dot{\tau}(P)=\int_{\partial B_2} H_{3j}(P,Q)n_j(P)F(Q)\,\mathrm{d}c_Q + \int_B H_{3j}(P,q)n_j(P)F^{(n)}(q)\,\mathrm{d}A_q - \int_{\partial B_1} H_{3j}(P,Q)n_j(P)D_K^{(n)}(Q)n_K(Q)\mathrm{d}c_Q \tag{9.14}$$

The boundary integrals in the above must, as usual, be interpreted in the sense of Cauchy principal values. It can be proved that H_{31} and H_{32}, with an arbitrary source point, are independent of the position of the field point provided it lies on the elliptical cutout. This makes the last terms in eqns. (9.11) and (9.14) considerably easier to evaluate.

9.2.4 Single Valued Displacements on Elliptic Cutout Boundary

When using the stress function approach, it is important to ensure that displacements are single valued on cutout boundaries in multiconnected regions. Here, the only nonzero displacement, u_3, must be single valued on the cutout boundary ∂B_1, i.e., it is required that

$$\int_{\partial B_1} \mathrm{d}u_3 = 0 \tag{9.15}$$

This equation can be written in terms of the stress function Φ by extending the analysis of Sokolnikoff[9] for the torsion of hollow beams, to this nonelastic case. The resulting equation is

$$\int_{\partial B_1} \frac{\mathrm{d}\dot{\Phi}}{\mathrm{d}n}\mathrm{d}c = \int_{\partial B_1} \mathbf{D}^{(n)}\cdot\mathbf{n}\mathrm{d}c \tag{9.16}$$

where $\mathbf{n}$ is the unit outward normal to the cutout surface (Fig. 9.2) and $D_1^{(n)}$ and $D_2^{(n)}$ are defined below eqn. (9.11).

The boundary element formulation presented in this chapter can be proved to satisfy eqn. (9.16) for an elliptic cutout. The proof is presented below for the case $a+b=2$ so that $m_1=(a-b)/2\equiv m$ and $m_2=1$. This makes $a=1+m$ and $b=1-m$. This is done for algebraic simplification and the proof goes through for the case $m_2 \neq 1$.

The normal derivative of $\dot{\Phi}$ at a point P on ∂B_1 (Fig. 9.2), from eqn.

(9.11), is

$$2\pi \frac{\mathrm{d}\dot{\Phi}}{\mathrm{d}n}(P)=2\pi(\dot{\sigma}_{31}n_2-\dot{\sigma}_{32}n_1)$$

$$=\int_{\partial B_2} \mathrm{Re}\left\{\frac{\mathrm{d}\Phi}{\mathrm{d}n}(P,Q)\right\}F(Q)\mathrm{d}c_Q$$

$$+\int_B \mathrm{Re}\left\{\frac{\mathrm{d}\phi}{\mathrm{d}n}(P,q)\right\}F^{(n)}(q)\,\mathrm{d}A_q$$

$$+\int_{\partial B_1} \mathrm{Re}\left\{\frac{\mathrm{d}\phi}{\mathrm{d}n}(P,Q)\right\}D_K^{(n)}(Q)n_K(Q)\mathrm{d}c_Q \qquad (9.17)$$

the positive sign on the last term being a consequence of the fact that $\mathbf{n}$ is the outward normal to ∂B_1. The normal derivatives of the kernel $\dot{\phi}$ are evaluated at the source point P on ∂B_1. There is no extra term due to a residue in this case.

The proof of eqn. (9.16) for single-valued displacements on ∂B_1 rests on the fact that, if the cutout is elliptic

$$\int_{\partial B_1} \mathrm{Re}\left\{\frac{\mathrm{d}\phi}{\mathrm{d}n}(P,q)\right\}\mathrm{d}c_P=\begin{cases} 0 \text{ if } q \text{ is outside } \partial B_1 \\ 2\pi \text{ if } Q \text{ is on } \partial B_1 \end{cases} \qquad (9.18)$$

Proof of eqn. (9.18) is sketched briefly below. The above result is also true in the limiting cases when the cutout is a circle or a line crack.

If a source point lies on the elliptical cutout ∂B_1

$$z=(1+m)\cos\alpha+\mathrm{i}(1-\mathrm{m})\sin\alpha$$

$$\xi=\mathrm{e}^{-\mathrm{i}\alpha}$$

where $0\leq\alpha\leq 2\pi$.

Let n_1 and n_2 be the components of the outward normal to ∂B_1 at some point on it. Then

$$\tilde{n}=n_1+\mathrm{i}n_2=\frac{\mathrm{e}^{\mathrm{i}\alpha}-m\mathrm{e}^{-\mathrm{i}\alpha}}{\sqrt{1+m^2-2m\cos 2\alpha}}$$

and a line element on the ellipse

$$\mathrm{d}c=\sqrt{1+m^2-2m\cos 2\alpha}\,\mathrm{d}\alpha$$

Let an arbitrary field point z_0 yield the root (see below eqn. (9.9))

$$r_i=\beta\mathrm{e}^{-\mathrm{i}\theta}\begin{cases} \beta=1 \text{ if } z_0 \text{ is on } \partial B_1 \\ \beta<1 \text{ if } z_0 \text{ is outside } \partial B_1 \end{cases}$$

Using all the above equations and eqn. (9.10) for $\phi(z, \bar{z}, z_0)$,

$$\mathrm{Re}\left\{\frac{\mathrm{d}\phi}{\mathrm{d}n}(P,q)\right\}\mathrm{d}c_P = \mathrm{Re}\left\{\frac{\partial\phi}{\partial z}\tilde{n} + \frac{\partial\phi}{\partial\bar{z}}\bar{\tilde{n}}\right\}\mathrm{d}c_P$$
$$= \frac{2(1-\delta\cos(\alpha-\theta))}{\delta^2 - 2\delta\cos(\alpha-\theta)+1}\mathrm{d}\alpha$$

where $\delta = 1/\beta \geq 1$, $\bar{\tilde{n}} = n_1 - in_2$ and P is on ∂B_1.

Now, with $\psi = \alpha - \theta$

$$\int_0^{2\pi} \frac{2(1-\delta\cos\psi)}{\delta^2 - 2\delta\cos\psi + 1}\mathrm{d}\psi = \begin{cases} 0 \text{ if } \delta > 1 (z_0 \text{ outside } \partial B_1) \\ 2\pi \text{ if } \delta = 1 \ (z_0 \text{ on } \partial B_1) \end{cases}$$

which proves eqn. (9.18).

Finally, integrating both sides of eqn. (9.17) around ∂B_1, and evaluating first the integrals around ∂B_1 in the resulting double integrals on the right-hand side of eqn. (9.17),

$$2\pi\int_{\partial B_1} \frac{\mathrm{d}\Phi}{\mathrm{d}n}\mathrm{d}c = 2\pi\int_{\partial B_1} D_K^{(n)} n_K \mathrm{d}c$$

which proves eqn. (9.16).

9.3 LINE INTEGRALS

Line integrals have played a very important role in both elastic and inelastic fracture mechanics. Path-independent line integrals are useful for determining conditions in a small region near a crack tip from remote measurements near a load line, and have been suggested as controlling parameters for crack growth. The best known path-independent line integral in fracture mechanics is the J integral due to Rice[10]

$$J = \int_\Gamma \left(Wn_1 - \tau_j \frac{\partial u_j}{\partial x_1}\right)\mathrm{d}c \qquad (9.19)$$

where $W = \int_0^\varepsilon \sigma_{ij}\mathrm{d}\varepsilon_{ij}$, u_j and τ_j are the displacement and traction components at a point on some path Γ and $n_1 = \mathbf{n}\cdot\mathbf{i}$ with $\mathbf{n}$ the unit outward normal to Γ at a point on it and $\mathbf{i}$ a unit vector along the x_1 axis. In linear elasticity, this integral is path-independent and W is the strain energy density. In fact, this is one of a class of path-independent integrals that can be derived systematically for linear elastic materials (see, for example, reference 11). The J integral is also path-independent in a deformation type elastic–plastic material where $d\varepsilon_{ij}$, in the definition of W, is the total (sum of elastic and plastic) strain increment.

An analogous integral suitable for the study of cracks in creeping plates was first defined by Landes and Begley.[12] This is the C integral defined as

$$C(t)=\int_{\Gamma}\left(W^{*}n_{1}-\tau_{j}\frac{\partial\dot{u}_{j}}{\partial x_{1}}\right)\mathrm{d}c \tag{9.20}$$

where $W^{*}=\int_{0}^{\dot{\varepsilon}}\sigma_{ij}\,\mathrm{d}\dot{\varepsilon}_{ij}$ can be interpreted as a rate potential. There is some question as to whether the increment of the total or the nonelastic strain rate should be used in the expression for W^{*}. In this section, $\mathrm{d}\dot{\varepsilon}_{ij}$ is taken to be the increment in the total strain rate, in a manner analogous to $\mathrm{d}\varepsilon_{ij}$ in the J integral. A similar integral can be defined, instead, with $\mathrm{d}\dot{\varepsilon}_{ij}^{(\mathrm{n})}$ in the expression for W^{*}, as has been done by Bassani and McClintock.[13] Such an integral is here called $\bar{C}(t)$. The two integrals C and $\bar{C}$ become identical in the limit of steady-state creep where zero stress rates lead to zero elastic strain rates, and this limiting value, called C^{*}, can be shown to be path-independent. This integral C^{*} has been shown to characterize the stress field near a crack tip in a plate undergoing steady creep deformation.[14] In the presence of elastic strain rates, however, as is the case, for example, during transient creep or with time-dependent loading, the integrals C and $\bar{C}$ have different values and both are path-dependent. Their values are obtained numerically for different paths, as functions of time, later in this chapter.

The special case of Mode III loading is of particular interest in this chapter. In this case, the two parts of the integrand in eqn. (9.20) assume simple forms. These, for an elastic-power-law creep material, are

$$\tau_{j}\frac{\partial\dot{u}_{j}}{\partial x_{1}}=2\{\sigma_{31}n_{1}+\sigma_{32}n_{2}\}\left\{\frac{\dot{\sigma}_{31}}{2G}+\dot{\varepsilon}_{31}^{(\mathrm{n})}\right\} \tag{9.21}$$

$$W^{*}=W_{1}^{*}+W_{2}^{*}$$

$$W_{1}^{*}=\int_{0}^{\dot{\varepsilon}}\sigma_{ij}\,\mathrm{d}\dot{\varepsilon}_{ij}^{(\mathrm{n})}=\frac{n}{n+1}\frac{\dot{\varepsilon}_{c}}{(\sigma_{c})^{n}}\sigma^{n+1} \tag{9.22}$$

$$W_{2}^{*}=\int_{0}^{\dot{\varepsilon}}\sigma_{ij}\,\mathrm{d}\dot{\varepsilon}_{ij}^{(\mathrm{e})}$$

$$=\frac{1}{G}\left\{(\sigma_{31}\dot{\sigma}_{31}+\sigma_{32}\dot{\sigma}_{32})\,\Big|_{0}^{t}-\int_{0}^{t}(\dot{\sigma}_{31}^{2}+\dot{\sigma}_{32}^{2})\,\mathrm{d}t'\right\} \tag{9.23}$$

The components of the outward unit normal to Γ are denoted by n_1 and n_2 in eqn. (9.21).

9.4 DISCRETIZATION AND IMPLEMENTATION OF BEM

9.4.1 Discretization of Integral Equations

The outer boundary of the body, ∂B_2, (Fig. 9.3) is divided into N_2 straight boundary elements using $N_b (N_b = N_2)$ boundary nodes and the inner boundary ∂B_1 into N_1 straight boundary segments. The interior of the body, B, is divided into n_i triangular internal elements. A discretized version of eqn. (9.14) is

$$\begin{aligned} 2\pi\dot{\tau}(P_M) = & \sum_{N_2} \int_{\Delta c_i} H_{3j}(P_M, Q) n_j(P_M) F(Q) \, dc_Q \\ & + \sum_{n_i} \int_{\Delta A_i} H_{3j}(P_M, q) n_j(P_M) F^{(n)}(q) \, dA_q \\ & - H_{3j}(P_M, \hat{Q}) n_j(P_M) \sum_{N_1} \int_{\Delta c_i} D_K^{(n)}(Q) n_K(Q) \, dc_Q \end{aligned} \tag{9.24}$$

where $\tau(P_M)$ are the traction components at the point P which coincides with the node M at the center of a segment on ∂B_2, Δc_i and ΔA_i are boundary and internal elements respectively and $\hat{Q}$ is *any* point on ∂B_1.

A very simple numerical scheme is used to obtain the numerical results presented in the next section. The concentrations F are assumed to be piecewise uniform on each boundary segment with their values assigned at the nodes which lie at the centers of each segment. The nonelastic strain rates $\dot{\varepsilon}_{3j}^{(n)}$ are interpolated linearly over each triangular internal cell so that $F^{(n)}$ is uniform within each cell. For the problems considered here, the last integral in eqn. (9.14) can be shown to vanish, i.e., in these cases

$$\int_{\partial B_1} \mathbf{D}^{(n)} \cdot \mathbf{n} \, dc = 0$$

by virtue of the stress pattern and constitutive model. Hence the last term in eqn. (9.24) is omitted in these calculations. In any case, in general, this term requires the evaluation of a *known* integral on ∂B_1 at each time step.

The integrals of H_{3j} on boundary elements are evaluated analytically for the singular and by Gaussian quadrature for the regular portions. Integrals of H_{3j} on triangular internal elements are obtained by Gaussian quadrature. This is adequate in these problems with the source points lying on the vertices of the triangles.

Substitution of the piecewise uniform source strengths into eqn. (9.24) and carrying out of the necessary integrations leads to an algebraic system of the type

$$\{\dot{\tau}\}=[A]\{F\}+\{d\} \tag{9.25}$$

The coefficients of the matrix $[A]$ contain boundary integrals of the kernels. The traction rates are prescribed, the vector $\{d\}$ contains the contributions from the area integral and the vector $\{F\}$ the unknown source strengths at the boundary nodes. As usual, the dimension of the unknown vector $\{F\}$ depends only on the number of boundary elements on ∂B_2.

Equation (9.11) for the stress rates at an internal point p is discretized in a similar fashion.

9.4.2 Determination of Line Integrals

The line integrals $C(t)$ and $\bar{C}(t)$ are evaluated at each time on three paths through eqns. (9.21)–(9.23). Each path is decomposed into a sufficient number of straight segments and the integral over each segment is obtained by Gaussian quadrature.

9.4.3 Solution Strategy

The solution strategy is as follows. The initial stress field is obtained by solving the corresponding elastic problem. The initial rates of the nonelastic strains are obtained from the constitutive model. The vector **d** in eqn. (9.25) is calculated next, and this, together with the prescribed rates of boundary tractions, is used to calculate the initial distribution of concentration F on ∂B_2. These concentrations are now used in the discretized version of eqn. (9.11) to calculate the initial stress rates throughout the body. These rates are used to determine the values of the variables after a small time interval Δt and so on, and in this way the time histories of the relevant variables are obtained. Once again, time integration is carried out by an Euler type stepwise procedure with automatic time-step control, as discussed in Section 4.2.1. The line integrals C and $\bar{C}$ are evaluated on each of the paths, at each time, through eqns. (9.21)–(9.23).

9.5 ANALYTICAL AND FINITE ELEMENT METHODS

9.5.1 Asymptotic Analytical Solutions for Line Cracks

Asymptotic solutions for stress redistribution near crack tips in creeping plates have been recently obtained by Riedel and Rice.[15–18] The plate material is elastic–nonlinear viscous, with the nonelastic strains obeying a power law creep constitutive model (eqns. (2.5), (2.8) and (2.9) of Section 2.1). The elastic strain rates in the vicinity of the crack tips are neglected relative to the nonelastic ones. The Mode III remote loads are either constant or increase in time at a constant rate. Only the dependence of stresses in a small region near a crack tip with the distance from the crack tip, 'ρ', and time, 't', are of interest.

If the remote loading is invariant in time, the stress field near the crack tip has the dependence[15–17]

$$\text{stress component } \alpha(\rho t)^{-1/(n+1)} \tag{9.26}$$

where 'ρ' is measured from the crack tip in the x_1 direction. The stress component of interest here is σ_{32} for Mode III.

If the remote loading increases linearly in time, the corresponding result, for Mode III is[18]

$$\text{stress concentration } \alpha(\rho t^n)^{-1/(n+1)} \tag{9.27}$$

in the region where the drop in the stress concentration takes place. The stress concentration is $\sigma_{32}/\sigma_{32}^\infty$. The quantity σ_{32}^∞ is some reference stress value in the work of Rice and Riedel. In the BEM analysis, it is, explicitly, the value of the remote stress.

The above results are also valid for Mode I loading where the stress component σ_{22} (loading in the 2 direction) and stress concentration $\sigma_{22}/\sigma_{22}^\infty$ are of interest.

The assumption of negligibly small elastic strain rates, relative to the nonelastic rates, is asymptotically correct as one approaches a crack tip, but may not be so even at a small distance away from it. Also, the existence of a steady-state stress distribution is impossible if the remote loading is time dependent.

9.5.2 Finite Element Solutions

The finite element method has been used to solve the complete transient problem in the presence of elastic strains by Ehlers and Riedel[19] and Bassani and McClintock.[13] In the interest of numerical efficiency and accuracy, the finite element method typically requires the use of special crack tip elements to reflect the proper singular behavior of stresses in this region. Thus, Bassani and McClintock, for example, impose a strain rate field which varies as $1/\rho$ near the crack tip. This approach is weak in the sense that the numerical method is unable to change from the zero time elastic singularity to the nonelastic stress distribution in a consistent, time-dependent fashion, and is not independent of the asymptotic analysis of the problem. As mentioned before, the boundary element method presented in this chapter requires no *a priori* assumptions about the stress distribution near a crack tip and satisfies the proper boundary conditions on the crack surface in an implicit manner.

9.6 NUMERICAL RESULTS BY BEM

9.6.1 Material Model and Loading

Material behavior is described by an elastic-power-law creep constitutive model. The appropriate equations for this model are (2.5), (2.8) and (2.9) of Section 2.1 with $\varepsilon^{(n)}=\varepsilon^{(c)}$. The material parameters representative of 304 stainless steel at 400° C, are[20]

$$G=9{\cdot}4\times 10^{6}\ \text{psi}$$

$$\dot{\varepsilon}_c=0{\cdot}277\times 10^{-3}\ \text{sec}^{-1}\ \text{at}\ \sigma_c=0{\cdot}18\times 10^{6}\ \text{psi}$$

$$n=7\ \ (n=2\ \text{is also used}).$$

Two kinds of Mode III remote loading σ_{32} are considered here. The first is time-independent remote stress and the second is remote stress increasing from zero at a constant rate.

9.6.2 Crack Model and Discretization

Center cracked square plates are considered here and a crack is modelled as a very thin ellipse with axis ratio 199 or 1999. A line crack is avoided since the nonelastic strain rates would become singular at a crack tip in such a case, and this would require an analytical description of $\dot{\varepsilon}_{31}^{(n)}$ and $\dot{\varepsilon}_{32}^{(n)}$ very near the crack tip in order to evaluate the area integrals in eqns. (9.11) and (9.14). The goal of the BEM solution is the numerical evaluation of the various quantities without any *a priori* analytical prescriptions regarding stress or strain fields.

The discretization used is shown in Fig. 9.4 and 9.5 for the cases $a/b=199$ and $a/b=1999$ respectively. Because of symmetry, only a quarter of a center-cracked plate needs to be considered. Further detail of the internal cells for the shaded portion in Fig. 9.5 is given in Fig. 9.16.

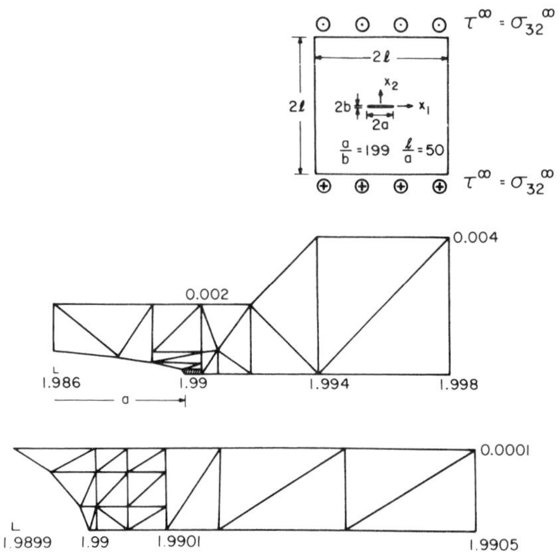

FIG. 9.4. Internal mesh for cracked plate with $a/b = 199$.

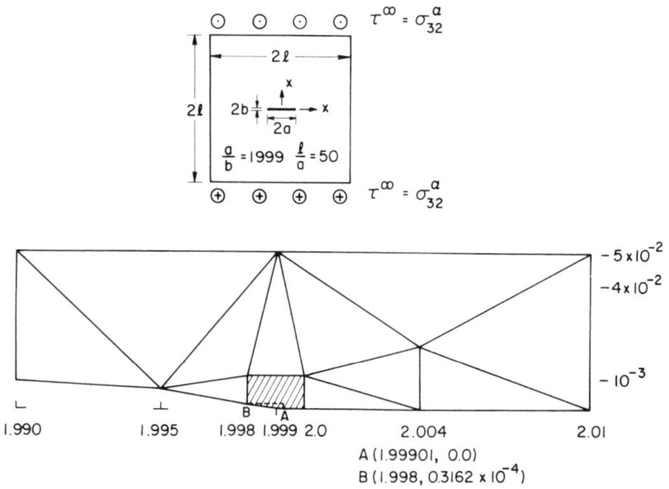

FIG. 9.5. Internal mesh for cracked plate with $a/b = 1999$. Small integration path AB. Details of shaded portion in Fig. 9.16.

Only a small part of the interior of the plate near the crack tip is divided into triangular cells. This is a particular advantage of the boundary element method in that the integrals over the body B in eqns. (9.11) and (9.14) involve known integrands at any time step and only that part of B

with significant values of nonelastic strain rate gradients (i.e. $F^{(n)}$) needs to be included. The choice of a mesh can be justified by checking these strain rates *a posteriori* at the final time of integration. Thus, in the numerical results discussed in this section, for example, the nonelastic strain rates (with $n=7$) at the outskirts of the mesh of Fig. 9.5, at the final time, are about nine orders of magnitude lower than the same quantities near the crack tip.

The boundary discretization uses 20 uniformly distributed boundary nodes along the plate edges in the first quadrant. The symmetry lines need not be discretized. The line integrals $C(t)$ and $\bar{C}(t)$ are numerically evaluated over the small (S), medium (M) and large (L) paths shown in Figs. 9.5 and 9.6. A line integral is evaluated on each straight piece of an integration path by using six Gauss points.

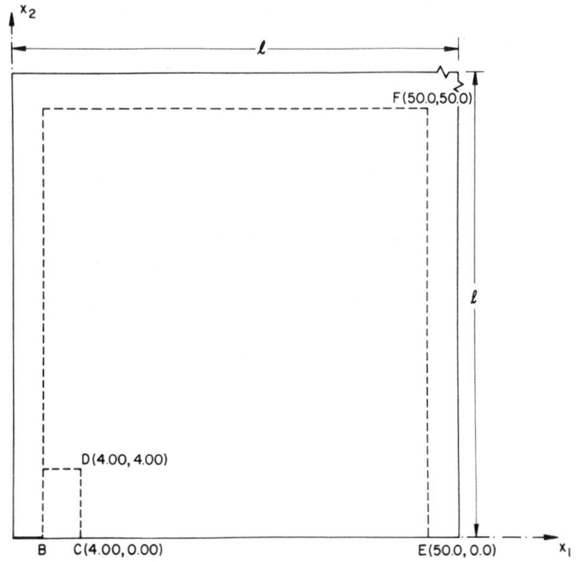

FIG. 9.6. Integration paths Medium (CDB) and Large (EFB). $a/b = 1\,999$.

9.6.3 Computer Program Verification

The BEM computer program has been verified by running two checks.

Elastic Stresses

An elastic solution has been obtained for Mode III loading of a center cracked plate with a line crack. This solution, with only eight boundary nodes placed uniformly along the plate edges in the first quadrant, has the proper square root singularity near the crack tip and gives a stress intensity factor within 5% of the analytical solution for an infinite plate with a crack.[21]

The J Integral

The J integral has been computed on the three paths shown in Figs. 9.5 and 9.6 for the elastic solution due to an applied remote stress of 1 ksi. The values on the three paths agree within about 0·1%.

Auxiliary Nonelastic Program HRR

In order to check further the algorithm for the numerical evaluation of the line integrals, an auxiliary program, called the HRR program, has been developed. The stress field near a crack tip in a creeping solid calculated by Riedel[15] (this is also called the HRR field for creep) has been input into this program and the integral

$$I(t)=\int_{\Gamma}\{W^*n_1-2\tau_3\dot{\varepsilon}_{31}^{(n)}\}\mathrm{d}c \qquad (9.28)$$

has been numerically evaluated *with the HRR field for creep* on several circular paths of different small radii centered at the right crack tip. This integral is path-independent even at short times since the HRR field neglects the elastic strain rates and thereby assumes the nonelastic strain rates to be compatible. This can be easily proved as follows:

$$W^*=2\int_0^{\dot{\varepsilon}}\{\sigma_{31}\mathrm{d}\dot{\varepsilon}_{31}^{(n)}+\sigma_{32}\mathrm{d}\dot{\varepsilon}_{32}^{(n)}\}$$

Using the power law creep constitutive model

$$\frac{\partial W^*}{\partial x_1}=2\left(\sigma_{31}\frac{\partial\dot{\varepsilon}_{31}^{(n)}}{\partial x_1}+\sigma_{32}\frac{\partial\dot{\varepsilon}_{32}^{(n)}}{\partial x_1}\right)$$

$$2\tau_3\dot{\varepsilon}_{31}^{(n)}=2(\sigma_{31}n_1+\sigma_{32}n_2)\dot{\varepsilon}_{31}^{(n)}$$

Now

$$\int_{\Gamma}\{W^*n_1-2\tau_3\dot{\varepsilon}_{31}^{(n)}\}\mathrm{d}c$$

$$=\int_A\left[\frac{\partial W^*}{\partial x_1}-2\left\{\frac{\partial}{\partial x_1}(\sigma_{31}\,\dot{\varepsilon}_{31}^{(n)})+\frac{\partial}{\partial x_2}(\sigma_{32}\,\dot{\varepsilon}_{31}^{(n)})\right\}\right]\mathrm{d}A$$

Using the equilibrium equation and the above formula for $\partial W^*/\partial x_1$, this becomes

$$2\int_A \left\{\sigma_{31}\frac{\partial \dot{\varepsilon}_{31}^{(n)}}{\partial x_1}+\sigma_{32}\frac{\partial \dot{\varepsilon}_{32}^{(n)}}{\partial x_1}-\left(\sigma_{31}\frac{\partial \dot{\varepsilon}_{31}^{(n)}}{\partial x_1}+\sigma_{32}\frac{\partial \dot{\varepsilon}_{31}^{(n)}}{\partial x_2}\right)\right\} \mathrm{d}A$$

which vanishes if the nonelastic strain rates are compatible, i.e., satisfy the compatibility equation

$$\frac{\partial \dot{\varepsilon}_{31}^{(n)}}{\partial x_2}=\frac{\partial \dot{\varepsilon}_{32}^{(n)}}{\partial x_1}$$

Unfortunately, of course, $I(t)$ is not path-independent if the complete transient solution is used since in that case the nonelastic strains are not, by themselves, compatible.

Numerically, the HRR computer program delivers the same value for $I(t)$ on several paths to six significant figures and also gives $I(t)\propto 1/t$, as expected. This further verifies the Gaussian integration logic used in the BEM program for the determination of the line integrals for the non-elastic problem.

9.6.4 Results for Stresses

Numerical results from the BEM computer program, for a cracked plate loaded in anti-plane shear (Fig. 9.4) are presented in Figs. 9.7–9.11. These results are taken from reference 7. Of these, Figs. 9.7–9.9 show the stress concentration at the crack tip as a function of time for various cases and the rest show the redistribution of stress near the crack tip along the x_1 axis. All the graphs are log-log plots.

Figure 9.7 shows typical plots for stress concentration at the crack tip

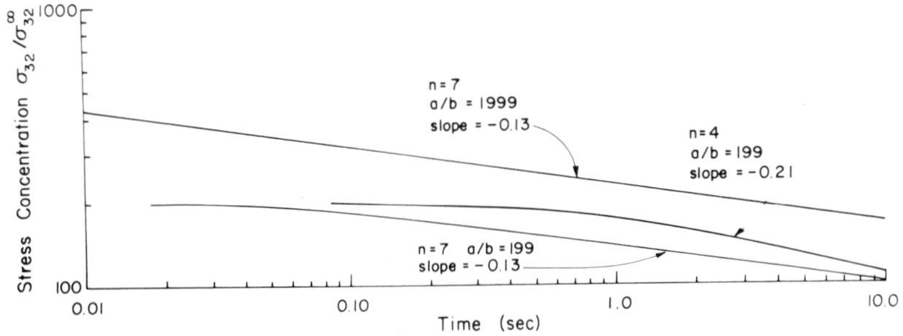

FIG. 9.7. Stress concentration at crack tip as a function of time for constant remote stress. $\sigma_{32}^{\infty}=1$ ksi.

versus time for constant remote stress. For $a/b = 199$, the stress concentration is seen to drop slowly in time for short times and the log-log plot becomes substantially straight beyond 1·5 seconds. The slope on these and other figures refers to the approximate slope of the straight portion of a curve as measured directly from the plots. The curve for a very thin ellipse ($a/b = 1999$) is seen to be substantially straight on this time scale.

Similar results for remote stress increasing at a constant rate are shown in Figs. 9.8 and 9.9. In these cases there is a certain time period before the stress concentration starts to decrease along nearly a straight line. At long times the concentration tends to level off in Fig. 9.8. This behavior is in general in qualitative agreement with Riedel.[18] Quantitative comparison of slopes is presented in Table 9.1 which will be discussed later in this section. Drop in stress concentration is due to flow of the material near the crack tip and consequent accommodation of stress.

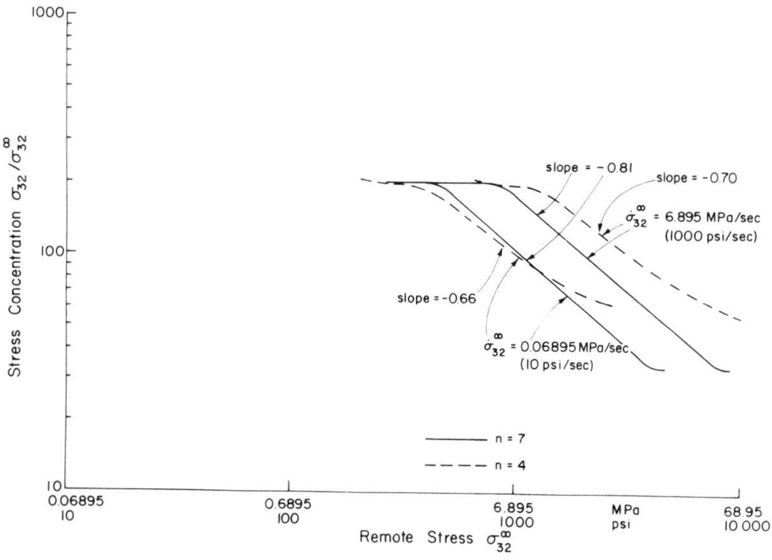

FIG. 9.8. Stress concentration at crack tip as a function of remote stress increasing at a constant rate. $a/b = 199$.

Redistribution of stress near the crack tip along the x_1 axis is shown in Figs. 9.10 and 9.11 for two values of a/b. The remote loading is constant in these cases. The elastic solution, as expected, is proportional to $\rho^{-1/2}$

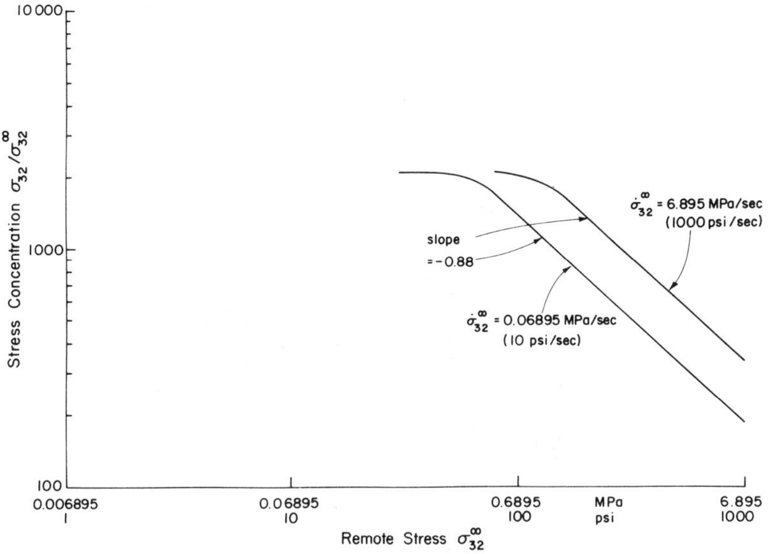

FIG. 9.9. Same situation as in Fig. 9.8. $n = 7$, $a/b = 1\,999$.

beyond a short distance from the crack tip. This distance is much less for the thinner crack (Fig. 9.11) compared to the other one (Fig. 9.10). The stress, of course, is finite at the crack tip in either case. The stress redistributes with time and becomes more and more weakly dependent on 'ρ' very near the crack tip but remains close to the elastic solution away from the crack. This, again, is in qualitative agreement with Riedel.[15] Quantitative agreement between the numerical and the asymptotic spatial stress field very near a crack tip is not expected to be good, since, in the numerical calculations, a crack is modelled as a very thin ellipse; and consequently has a large but finite stress at its tips.

The numerically calculated dependence of the stress concentration at the crack tip, with time, is compared with asymptotic analytical results[15–18] for various cases in Table 9.1. As mentioned before, these asymptotic results are for line cracks in plates where the elastic strain rates near the crack tip are neglected relative to the nonelastic ones. The numerical slopes refer to approximate slopes of the straight portions of the appropriate curves, measured directly from the plots. The correlation of slopes, in general, seems to improve with increase of axis ratio a/b and power law index 'n'. This is reasonable in view of the assumptions used in the asymptotic solution.

TABLE 9.1
DEPENDENCE OF STRESS CONCENTRATION AT CRACK TIP WITH TIME FOR MODE III

Figure	*a/b*	*Loading*	*n*	*Slope (analytical)*	*Slope (numerical)*	*Slope* $\left(\frac{analyt.-num.}{analyt.}\right)\times 100$
9·7	199	$\sigma_{32}^{\infty}=1$ ksi	7	$-1/(n+1)=-0{\cdot}125$	−0·13	−4
9·7	1 999	$\sigma_{32}^{\infty}=1$ ksi	7	−0·125	−0·13	−4
9·7	199	$\sigma_{32}^{\infty}=1$ ksi	4	−0·20	−0·21	−5
9·8	199	$\dot{\sigma}_{32}^{\infty}=1$ ksi/sec	7	$-n/(n+1)=-0{\cdot}875$	−0·81	7·4
9·8	199	$\dot{\sigma}_{32}^{\infty}=10$ psi/sec	7	−0·875	−0·81	8
9·8	199	$\dot{\sigma}_{32}^{\infty}=1$ ksi/sec	4	−0·80	−0·70	12·5
9·8	199	$\dot{\sigma}_{32}^{\infty}=10$ psi/sec	4	−0·80	−0·66	17·5
9·9	1 999	$\dot{\sigma}_{32}^{\infty}=1$ ksi/sec	7	−0·875	−0·88	−0·57
9·9	1 999	$\dot{\sigma}_{32}^{\infty}=10$ psi/sec	7	−0·875	−0·88	−0·57

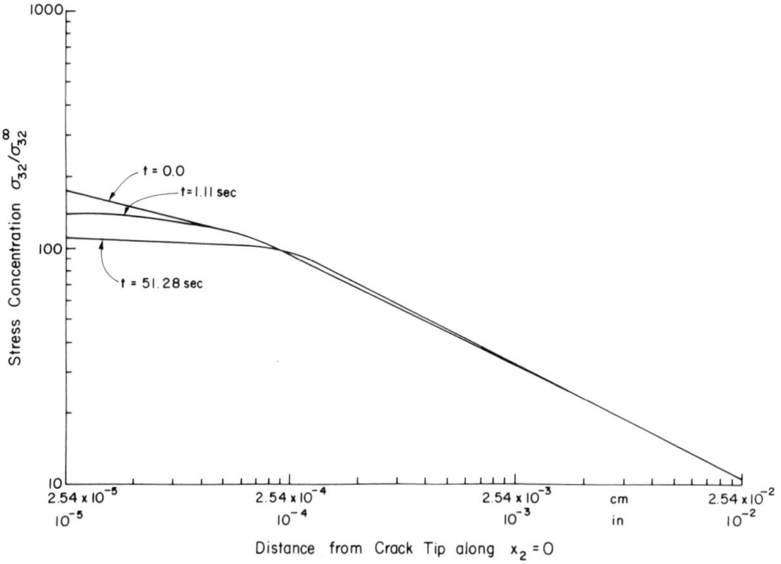

FIG. 9.10. Stress concentration near crack tip for constant remote stress $\sigma_{32}^{\infty}=1$ ksi. $n=7$, $a/b=199$.

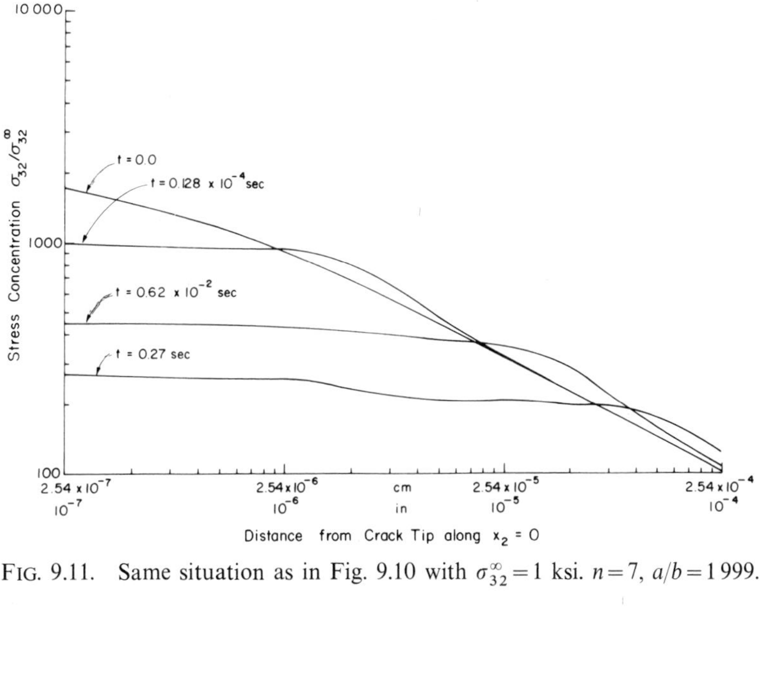

FIG. 9.11. Same situation as in Fig. 9.10 with $\sigma_{32}^{\infty}=1$ ksi. $n=7$, $a/b=1999$.

9.6.5 Line Integral Calculations

Further Results from the HRR Program

The HRR program, described earlier, is also used to estimate how long it takes for the $\bar{C}$ integral to become path independent in a certain region. Table 9.2[22] gives the values of the $\bar{C}$ integral, based on the HRR field, on four circular paths centered at the right crack tip at five different times. It is found as expected that the values of the integral on the different paths converge with increasing time. This is, of course, related to the magnitude of $\dot{\varepsilon}_{32}^{(e)}$ relative to $\dot{\varepsilon}_{32}^{(n)}$. The smaller the value of the elastic strain rate relative to the nonelastic one, the closer the values of the $\bar{C}$ integrals on different paths. Thus, for example, at the shortest time ($0{\cdot}1 \times 10^{-6}$ sec), when the elastic strain rates are large, the $\bar{C}(t)$ integral is substantially different even on the two smallest paths.

Line Integrals for Time-independent Remote Loading

The full BEM computer program has been used to solve the complete transient problem including elastic strain rates and the numerically calculated $C(t)$ and $\bar{C}(t)$ integrals, on three different paths, are shown in Figs. 9.12 and 9.13 respectively.[22] It is seen that the integrals approach a small value at 1000 seconds, and these values on the medium and large paths are of the same order as C^* for a compact tension specimen. The values of the integrals, at short times, are very much path dependent. At short times, the integrals have large values on the small path but relatively small values on the medium and large ones since the stresses and stress rates are initially small away from the immediate neighborhood of the crack tips. Further, while $C(t)$ decreases with time, as expected, on all the paths, the time dependence of $\bar{C}(t)$ with time is qualitatively different on the three paths.

Line Integrals for Remote Loading Increasing at a Uniform Rate

The computed values of the C and $\bar{C}$ integrals, on the three paths, at different times, are shown in Tables 9.3 and 9.4.[23] Two values of the creep exponent n are used in the two cases. The remote loading rate is $\dot{\sigma}_{32}^{\infty} = 1$ ksi/sec with $\sigma_{32}^{\infty}(0) = 0$ in each case. Both the C and $\bar{C}$ integrals are seen to be path dependent. This is due to the existence of large values of elastic strain rates at short times. In fact, a calculation with a suddenly applied load of $\sigma_{32}^{\infty} = 1$ ksi, for example, yields strain rates $\dot{\varepsilon}_{32}^{(e)} = 2905$ sec^{-1} and $\dot{\varepsilon}_{32}^{(n)} = 1843$ sec^{-1} at the crack tip at time 0^{+} (see also the HRR results in Table 9.2). The comparable magnitude of the two strain rates

TABLE 9.2
INTEGRALS FROM HRR FIELD. $n=7$

t sec	$\dot{\varepsilon}_{32}^{(e)}$ sec^{-1}	$\dot{\varepsilon}_{32}^{(n)}$ sec^{-1}	$\bar{C}$ lb/in. sec	Path radius r in.
	$0{\cdot}225\,123\,69 \times 10^{6}$	$0{\cdot}225\,346\,05 \times 10^{6}$	$0{\cdot}359\,547\,9 \times 10^{7}$	$0{\cdot}1 \times 10^{-6}$
	$0{\cdot}126\,596\,35 \times 10^{6}$	$0{\cdot}400\,728\,24 \times 10^{6}$	$0{\cdot}150\,489\,39 \times 10^{7}$	$0{\cdot}1 \times 10^{-4}$
$0{\cdot}1 \times 10^{-6}$	$0{\cdot}949\,338\,71 \times 10^{5}$	$0{\cdot}534\,379\,70 \times 10^{3}$	$-0{\cdot}847\,638\,02 \times 10^{7}$	$0{\cdot}1 \times 10^{-3}$
	$0{\cdot}533\,852\,39 \times 10^{5}$	$0{\cdot}950\,276\,41 \times 10$	$-0{\cdot}380\,240\,81 \times 10^{9}$	$0{\cdot}1 \times 10^{-1}$
	$0{\cdot}126\,596\,35 \times 10^{4}$	$0{\cdot}400\,728\,24 \times 10^{4}$	$0{\cdot}364\,215\,94 \times 10^{5}$	$0{\cdot}1 \times 10^{-6}$
	$0{\cdot}711\,903\,60 \times 10^{3}$	$0{\cdot}712\,606\,78 \times 10^{2}$	$0{\cdot}298\,105\,83 \times 10^{5}$	$0{\cdot}1 \times 10^{-4}$
$0{\cdot}1 \times 10^{-4}$	$0{\cdot}533\,852\,39 \times 10^{3}$	$0{\cdot}950\,276\,41 \times 10$	$-0{\cdot}175\,297\,67 \times 10^{4}$	$0{\cdot}1 \times 10^{-3}$
	$0{\cdot}300\,207\,26 \times 10^{3}$	$0{\cdot}168\,985\,70$	$-0{\cdot}117\,737\,53 \times 10^{7}$	$0{\cdot}1 \times 10^{-1}$
	$0{\cdot}711\,903\,60 \times 10$	$0{\cdot}712\,606\,78 \times 10^{2}$	$0{\cdot}365\,692\,1 \times 10^{3}$	$0{\cdot}1 \times 10^{-6}$
	$0{\cdot}400\,332\,81 \times 10$	$0{\cdot}126\,721\,40 \times 10$	$0{\cdot}344\,786\,25 \times 10^{3}$	$0{\cdot}1 \times 10^{-4}$
$0{\cdot}1 \times 10^{-2}$	$0{\cdot}300\,207\,26 \times 10$	$0{\cdot}168\,985\,70$	$0{\cdot}244\,973\,51 \times 10^{3}$	$0{\cdot}1 \times 10^{-3}$
	$0{\cdot}168\,818\,95 \times 10$	$0{\cdot}300\,503\,79 \times 10^{-2}$	$-0{\cdot}347\,267\,08 \times 10^{4}$	$0{\cdot}1 \times 10^{-1}$
	$0{\cdot}300\,207\,26 \times 10^{-2}$	$0{\cdot}168\,985\,7$	$0{\cdot}366\,253\,39$	$0{\cdot}1 \times 10^{-6}$
	$0{\cdot}168\,818\,95 \times 10^{-2}$	$0{\cdot}300\,503\,79 \times 10^{-2}$	$0{\cdot}362\,535\,75$	$0{\cdot}1 \times 10^{-4}$
$1{\cdot}0$	$0{\cdot}126\,596\,35 \times 10^{-2}$	$0{\cdot}400\,728\,24 \times 10^{-3}$	$0{\cdot}344\,786\,25$	$0{\cdot}1 \times 10^{-3}$
	$0{\cdot}711\,903\,60 \times 10^{-3}$	$0{\cdot}712\,606\,78 \times 10^{-5}$	$-0{\cdot}316\,314\,78$	$0{\cdot}1 \times 10^{-1}$
	$0{\cdot}126\,596\,35 \times 10^{-5}$	$0{\cdot}400\,728\,24 \times 10^{-3}$	$0{\cdot}366\,353\,2 \times 10^{-3}$	$0{\cdot}1 \times 10^{-6}$
	$0{\cdot}711\,903\,60 \times 10^{-6}$	$0{\cdot}712\,606\,78 \times 10^{-5}$	$0{\cdot}365\,692\,1 \times 10^{-3}$	$0{\cdot}1 \times 10^{-4}$
$1\,000{\cdot}0$	$0{\cdot}533\,852\,39 \times 10^{-6}$	$0{\cdot}950\,276\,41 \times 10^{-6}$	$0{\cdot}362\,535\,75 \times 10^{-3}$	$0{\cdot}1 \times 10^{-3}$
	$0{\cdot}300\,207\,26 \times 10^{-6}$	$0{\cdot}168\,985\,70 \times 10^{-7}$	$0{\cdot}244\,973\,51 \times 10^{-3}$	$0{\cdot}1 \times 10^{-1}$

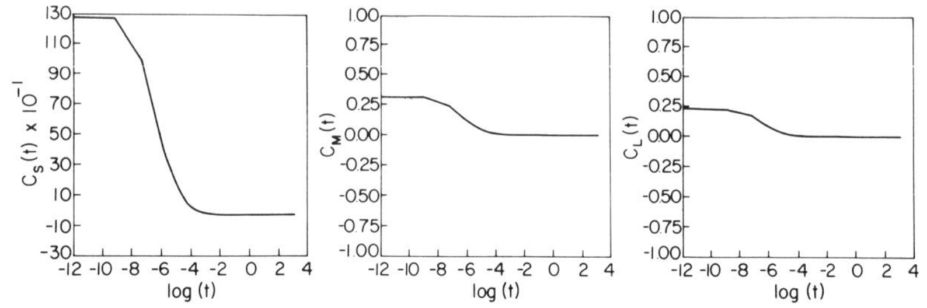

FIG. 9.12. Variation of $C(t)$ with log t on three paths. $\sigma_{32}^{\infty}=1\,000$ psi, $n=7{\cdot}0$, $a/b=1\,999$, time in seconds.

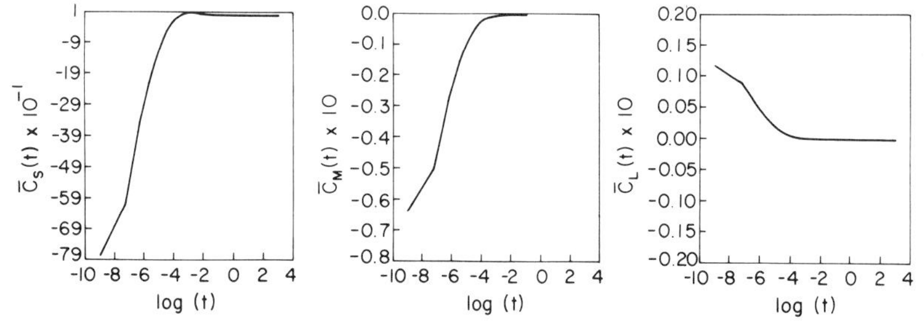

FIG. 9.13. Variation of $\bar{C}(t)$ with log t on three paths. $\sigma_{32}^{\infty}=1\,000$ psi, $n=7{\cdot}0$, $a/b=1\,999$, time in seconds.

makes C and $\bar{C}$ strongly path-dependent at short times, even for small paths near the crack tips.

The integrals C and $\bar{C}$ differ in the choice of W^* but otherwise use the same quantities. This, in fact, leads to very nearly equal values of C and $\bar{C}$ over the medium and large paths where the contribution of W^* in C is very small. The integrals over the smallest path start diverging fairly quickly as soon as the contribution of W^* in C starts becoming significant. This difference is much more pronounced for $n=7$ than for $n=2$.

The overall variation of the dimensionless C integral on the different paths, with time, is shown in Figs. 9.14 and 9.15 for two values of n. The C integral is defined as

$$\hat{C}(t)=\frac{C(t)}{\sigma_{\infty}\dot{\varepsilon}_{\infty}^{(n)}a} \tag{9.29}$$

where σ_{∞} and $\dot{\varepsilon}_{\infty}^{(n)}$ are the remote stress and nonelastic strain rate invariants, respectively, and a is half the crack length. This integral decreases in time owing to the strong increase of the denominator in eqn.

TABLE 9.3

COMPARISON OF $C(t)$ AND $\bar{C}(t)$ FOR DIFFERENT PATHS. $n=2$, $a/b=1\,999$

Time	C_s(*lb/in. sec*)	$\bar{C}_s$(*lb/in. sec*)	C_M(*lb/in. sec*)	$\bar{C}_M$(*lb/in. sec*)	C_L(*lb/in. sec*)	$\bar{C}_L$(*lb/in. sec*)
1×10^{-5}	$1{\cdot}414 \times 10^{-6}$	$1{\cdot}414 \times 10^{-6}$	$4{\cdot}63 \times 10^{-6}$	$4{\cdot}63 \times 10^{-6}$	$1{\cdot}69 \times 10^{-6}$	$1{\cdot}69 \times 10^{-6}$
0·57	$8{\cdot}47 \times 10^{-2}$	$8{\cdot}49 \times 10^{-2}$	0·266	0·266	$9{\cdot}69 \times 10^{-2}$	$9{\cdot}69 \times 10^{-2}$
1·19	0·197 5	0·201 6	0·55	0·55	0·201 1	0·201 1
2·91	0·76	0·886	1·35	1·349	0·496 9	0·496 3

TABLE 9.4

COMPARISON OF $C(t)$ AND $\bar{C}(t)$ FOR DIFFERENT PATHS. $n=7$, $a/b=1\,999$

Time	C_s(*lb/in. sec*)	$\bar{C}_s$(*lb/in. sec*)	C_M(*lb/in. sec*)	$\bar{C}_M$(*lb/in. sec*)	C_L(*lb/in. sec*)	$\bar{C}_L$(*lb/in. sec*)
1×10^{-5}	$1{\cdot}414 \times 10^{-6}$	$1{\cdot}414 \times 10^{-6}$	$4{\cdot}63 \times 10^{-6}$	$4{\cdot}63 \times 10^{-6}$	$1{\cdot}69 \times 10^{-6}$	$1{\cdot}69 \times 10^{-6}$
0·60	0·144 3	0·193 3	0·279 7	0·279 7	0·102 1	0·102 1
1·23	0·529 3	0·862 3	0·569 2	0·568 8	0·207 9	0·207 7
1·82	0·755 6	1·139 6	0·844 6	0·843 6	0·308 8	0·308 2
2·24	1·182 6	1·471 7	1·039 6	1·037 6	0·380 3	0·379 2

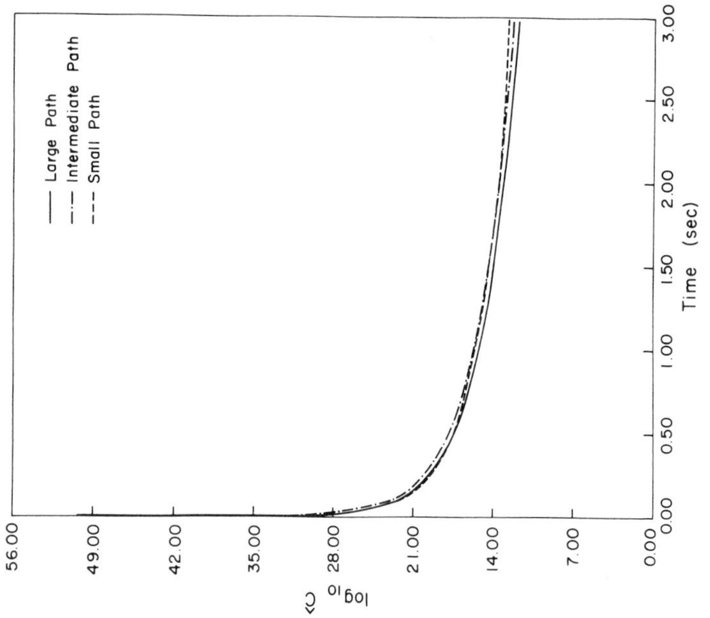

FIG. 9.14. Dimensionless C integral as function of time over three paths. $\dot{\sigma}_{32}^{\infty} = 1$ ksi/sec. $n = 2{\cdot}0$, $a/b = 1999$.

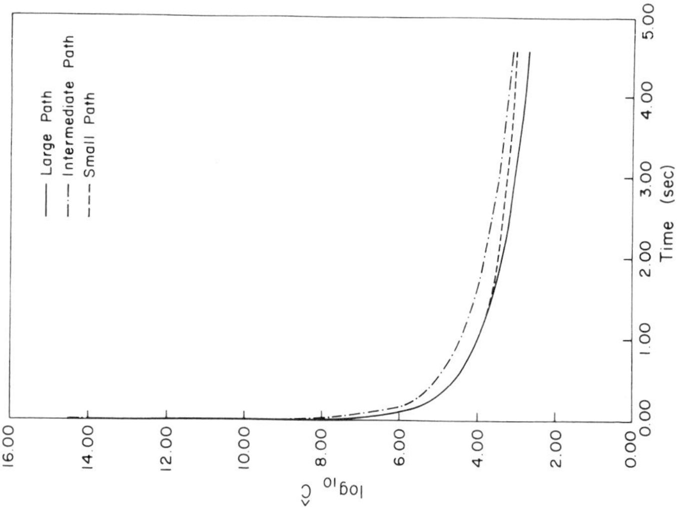

FIG. 9.15. Same situation as in Fig. 9.14 with $n = 7$.

(9.29) as a function of time. The overall qualitative behavior of $\hat{C}$ is seen to be similar in all cases and shows the expected relaxation with time.

Growth of Creep Zone

The boundary of the creep zone is defined in the usual way as the locus of points where the elastic and nonelastic strain invariants are equal, i.e. $\varepsilon^{(e)} = \varepsilon^{(n)}$. The growth of the creep zone for the problem discussed in the preceding section is shown in Fig. 9.16 for the case $n=7$. The creep zones are seen to be nearly circular as has been suggested by Riedel.[15]

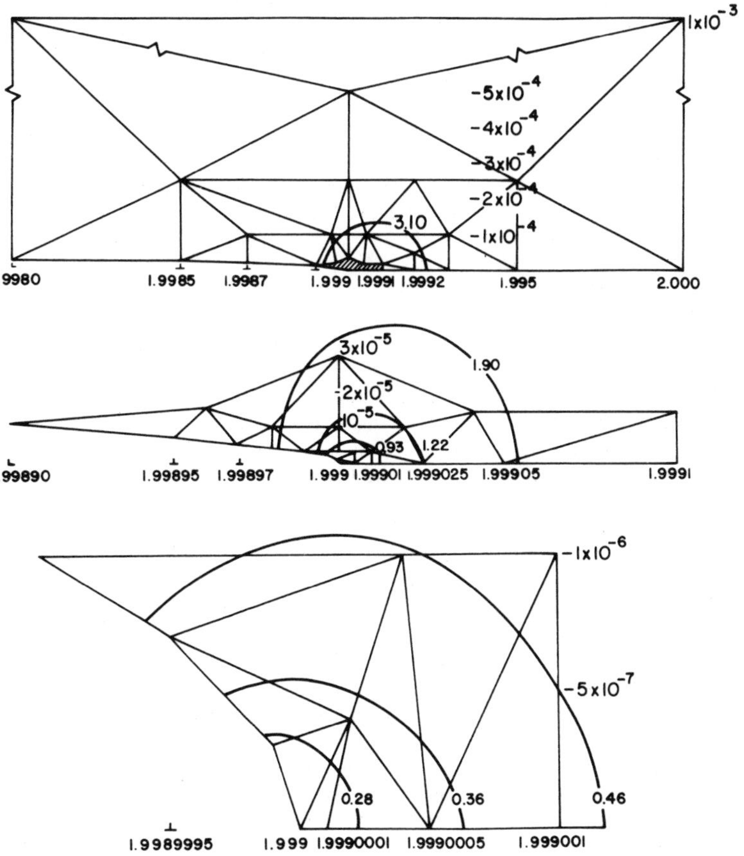

FIG. 9.16. Creep zones as functions of time. $a/b = 1999$.

REFERENCES

1. Cruse, T. A. Two dimensional b.i.e. fracture mechanics analysis. *Recent Advances in Boundary Element Methods*, C. A. Brebbia (ed.), Pentech Press, London and Plymouth, 167–184 (1978).
2. Mir-Mohamad-Sadegh, A. and Altiero, N. J. A boundary-integral approach to the problem of an elastic region weakened by an arbitrarily shaped hole. *Mechanics Research Communications*, **6**, 167–175 (1979).
3. Mir-Mohamad-Sadegh, A. and Altiero, N. J. Solution of the problem of a crack in a finite region using an indirect boundary-integral method. *Engineering Fracture Mechanics*, **11**, 831–837 (1979).
4. Mukherjee, S. and Morjaria, M. A boundary element formulation for planar time-dependent inelastic deformation of plates with cutouts. *International Journal of Solids and Structures*, **17**, 115–126 (1981).
5. Morjaria, M. and Mukherjee, S. Numerical analysis of planar, time-dependent inelastic deformation of plates with cracks by the boundary element method. *International Journal of Solids and Structures*, **17**, 127–143 (1981).
6. Mukherjee, S. and Morjaria, M. Boundary element analysis of time-dependent inelastic deformation of cracked plates loaded in antiplane shear. *International Journal of Solids and Structures*, **17**, 753–763 (1981).
7. Morjaria, M. and Mukherjee, S. Numerical solutions for stresses near crack tips in time-dependent inelastic fracture mechanics. *International Journal of Fracture*, **18**, 293–310 (1982).
8. Milne-Thomson, L. M. *Theoretical Aerodynamics* (4th ed.). Macmillan, New York (1966).
9. Sokolnikoff, I. S. *Mathematical Theory of Elasticity*. McGraw-Hill, New York (1956).
10. Rice, J. R. A path independent integral and the approximate analysis of strain concentration by notches and cracks. *Journal of Applied Mechanics*, American Society of Mechanical Engineers, **35**, 379–386 (1968).
11. Delph, T. J. Conservation laws in linear elasticity based upon divergence transformations. *Journal of Elasticity* (in press).
12. Landes, J. D. and Begley, J. A. A fracture mechanics approach to creep crack growth. American Society for Testing and Materials, *Special Technical Publication* 590, 128–148 (1976).
13. Bassani, J. L. and McClintock, F. A. Creep relaxation of stress around a crack tip. *International Journal of Solids and Structures*, **17**, 479–492 (1981).
14. Kumar, V. and Shih, C. F. Fully plastic crack solutions with applications to creep crack growth. *Proceedings of the Second International Conference on Engineering Aspects of Creep*, Institution of Mechanical Engineers, London, England, 211–214 (1980).
15. Riedel, H. Cracks loaded in anti-plane shear under creep conditions. *Zeitschrift für Metallkunde*, **69**, 755–760 (1978).
16. Riedel, H. and Rice, J. R. Tensile cracks in creeping solids. American Society for Testing and Materials. *Special Technical Publication* 700, 112–130 (1980).

17. RIEDEL, H. *Creep Deformation at Crack Tips in Elastic-Viscoplastic Solids.* Report No. MRL-E-114, Materials Research Laboratory, Brown University (1979).
18. RIEDEL, H. Private communication.
19. EHLERS, R. and RIEDEL, H. A finite element analysis of creep deformation in a specimen containing a macroscopic crack. *Presented at ICF 5, the 5th International Conference on Fracture*, Cannes, France (1981).
20. ODQUIST, F. K. G. *Mathematical Theory of Creep and Creep Rupture.* Clarendon Press, Oxford, UK (1966).
21. BENTHEM, J. P. and KOITER, W. T. Asymptotic approximations to crack problems, in *Mechanics of Fracture* 1, G. C. Sih (ed.), Noordhoff, Leyden, The Netherlands, 131–178 (1973).
22. BANTHIA, V. and MUKHERJEE, S. On stresses and line integrals in the presence of cracks. *Res Mechanica* (in press).
23. BANTHIA, V. and MUKHERJEE, S. Boundary element analysis of stresses in a creeping plate with a crack. *American Society for Testing and Materials. Special Technical Publication* (in press).

CHAPTER 10

Planar Problems of Inelastic Fracture

The boundary element method with augmented kernels, discussed in the previous chapter, is applied here to planar problems of inelastic deformation in the presence of cracks. A stress function formulation gives rise to a nonhomogeneous biharmonic equation for the stress function rate in this case. This equation is solved by the boundary element method. Numerical results for several sample problems are presented and compared with solutions obtained by asymptotic methods.

10.1 GOVERNING DIFFERENTIAL EQUATIONS

A planar body is considered with the x_1 and x_2 axes in the plane of the body and the x_3 axis normal to it. A stress function Φ is defined in the usual way

$$\sigma_{11}=\frac{\partial^2\Phi}{\partial x_2^2},\quad \sigma_{22}=\frac{\partial^2\Phi}{\partial x_1^2},\quad \sigma_{12}=-\frac{\partial^2\Phi}{\partial x_1\partial x_2} \tag{10.1}$$

where σ_{11}, σ_{22} and σ_{12} are the stress components.

The strain rates are decomposed, as usual, into elastic and nonelastic components. Using Hooke's law to relate the rates of elastic strains and stresses, and the compatibility equation in rate form

$$\frac{\partial^2\dot{\varepsilon}_{11}^{(e)}}{\partial x_2^2}+\frac{\partial^2\dot{\varepsilon}_{22}^{(e)}}{\partial x_1^2}-2\frac{\partial^2\dot{\varepsilon}_{12}^{(e)}}{\partial x_1\partial x_2}=-\left[\frac{\partial^2\dot{\varepsilon}_{11}^{(n)}}{\partial x_2^2}+\frac{\partial^2\dot{\varepsilon}_{22}^{(n)}}{\partial x_1^2}-2\frac{\partial^2\dot{\varepsilon}_{12}^{(n)}}{\partial x_1\partial x_2}\right] \tag{10.2}$$

results in an inhomogeneous biharmonic equation for the rate of the stress function

$$\nabla^4\dot{\Phi}=C^{(n)} \tag{10.3}$$

The function $C^{(n)}$ has the forms

$$C^{(n)} = -E\left[\frac{\partial^2\dot{\varepsilon}_{11}^{(n)}}{\partial x_2^2} + \frac{\partial^2\dot{\varepsilon}_{22}^{(n)}}{\partial x_1^2} - 2\frac{\partial^2\dot{\varepsilon}_{12}^{(n)}}{\partial x_1 \partial x_2}\right] \text{ for plane stress } (\sigma_{33}=0)$$

$$C^{(n)} = -\frac{E}{1-\nu^2}\left[\frac{\partial^2\dot{\varepsilon}_{11}^{(n)}}{\partial x_2^2} + \frac{\partial^2\dot{\varepsilon}_{22}^{(n)}}{\partial x_1^2} - \frac{2\partial^2\dot{\varepsilon}_{12}^{(n)}}{\partial x_1 \partial x_2} + \nu\nabla^2(\dot{\varepsilon}_{11}^{(n)} + \dot{\varepsilon}_{22}^{(n)})\right] \text{ for plane strain } (\varepsilon_{33}=0).$$

The admissible boundary conditions for the problems considered in this chapter are prescribed histories of traction on the outside boundary of the body.

10.2 BOUNDARY ELEMENT FORMULATION

10.2.1 Simply Connected Body

The biharmonic equation (10.3) can be transformed into an integral equation in several ways. An indirect formulation is presented here with two fundamental solutions of the equation, $r^2 \ln r$ and $\partial/\partial n_Q\,(r^2 \ln r)$ (see Section 8.2.2 regarding the choice of fundamental solutions). This gives the equation

$$\begin{aligned}8\pi\dot{\Phi}(p) &= \int_{\partial B} (r^2 \ln r)_{pQ} C_1(Q) \mathrm{d}c_Q \\ &+ \int_{\partial B} \frac{\partial}{\partial n_Q}(r^2 \ln r)_{pQ} C_2(Q) \mathrm{d}c_Q \\ &+ \int_B (r^2 \ln r)_{pq} C^{(n)}(q) \mathrm{d}A_q \qquad (10.4)\end{aligned}$$

Here C_1 and C_2 are unknown concentration functions to be determined from boundary conditions. Two functions are necessary for the biharmonic equation.

It is convenient to rewrite this equation in terms of complex variables with a view towards conformal mapping techniques that will be used later to derive the necessary augmented kernels. This has the form (see Fig. 9.1)

$$\begin{aligned}8\pi\dot{\Phi}(p) &= \int_{\partial B} K_1(p,Q) C_1(Q) \mathrm{d}c_Q + \int_{\partial B} K_2(p,Q) C_2(Q) \mathrm{d}c_Q \\ &+ \int_B K_1(p,q) C^{(n)}(q) \mathrm{d}A_q \qquad (10.5)\end{aligned}$$

where

$$K_1 = \mathrm{Re}\{\bar{z}\hat{\phi}_1(z, z_0) + \hat{\chi}_1(z, z_0)\}$$
$$K_2 = \mathrm{Re}\{\bar{z}\hat{\phi}_2(z, z_0) + \hat{\chi}_2(z, z_0)\}$$

with

$$\hat{\phi}_1(z, z_0) = (z - z_0)\ln(z - z_0)$$
$$\hat{\chi}_1(z, z_0) = -\bar{z}_0(z - z_0)\ln(z - z_0)$$
$$\hat{\phi}_2(z, z_0) = \frac{\partial}{\partial n_0}\hat{\phi}_1(z, z_0),\ \hat{\chi}_2(z, z_0) = \frac{\partial}{\partial n_0}\hat{\chi}_1(z, z_0)$$

where $\mathbf{n}_0$ is the outward normal at the field point Z_0 on the boundary ∂B, and, as usual, Re denotes the real part of the complex function within brackets and a superscribed bar denotes the complex conjugate.

The primary interest in the problems is the determination of stresses rather than the stress function. Thus, it is convenient to write the corresponding equations for stress rates. Also, when the traction equations are written for a point Z on the boundary, care must be taken to include residues, if any, that are generated from the singular kernels. These matters are given careful attention in the next section when the equations for multiply connected regions using augmented kernels are presented.

10.2.2 Body with Cutout

The singular kernels K_1 and K_2 in eqn. (10.5) are augmented with regular kernels so that the sum of these guarantees a traction-free inner boundary in a body with cutout. The new kernels are derived by using the methods of Muskhelishvili.[1] The approach is similar to that used by Sadegh and Altiero[2,3] for the analogous elastic problem and will be briefly outlined here.

An infinite plane with a cutout of contour ∂B_1 is shown in Fig. 10.1. The traction resultants F_1 and F_2 on a portion of arc AB on ∂B_1 due to the functions $\hat{\phi}$ and $\hat{\chi}$ are[1,4] ($\hat{\phi}$ can be either $\hat{\phi}_1$ or $\hat{\phi}_2$ of eqn. (10.5) and similarly for $\hat{\chi}$)

$$F_1 + \mathrm{i}F_2 = \int_A^B (\tau_1 + \mathrm{i}\tau_2)\mathrm{d}c$$
$$= -\mathrm{i}[\hat{\phi}(Z, z_0) + Z\overline{\hat{\phi}'(Z, z_0)} + \overline{\hat{\psi}(Z, z_0)}]_A^B \quad (10.6)$$

where τ_1 and τ_2 are the components of traction at a point Z in AB on ∂B_1, dc is an element of the curve ∂B_1 and $\psi(z, z_0) = \chi'(z, z_0)$, the prime

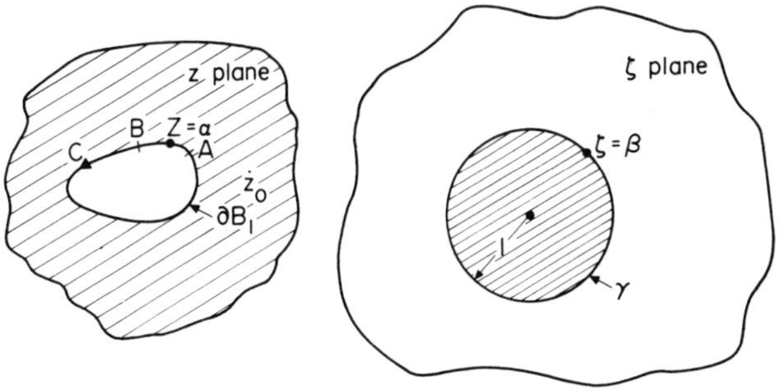

FIG. 10.1.

denoting differentiation with respect to the variable argument z. If a mapping function

$$z = w(\xi) \tag{10.7}$$

can be found which maps the region on and outside ∂B_1 in the z plane to a region on and inside a unit circle γ in the ξ plane, the expression within brackets on the right-hand side of eqn. (10.6) at a point $Z=\alpha$ on ∂B_1 can be written as

$$-F(\beta, z_0) \equiv \hat{f}(\beta, z_0) + \frac{w(\beta)}{\overline{w'(\beta)}} \overline{\hat{f}'(\beta, z_0)} + \overline{\hat{g}(\beta, z_0)} \tag{10.8}$$

where
$$\hat{\phi}(z, z_0) = \hat{f}(\xi, z_0)$$
$$\hat{\psi}(z, z_0) = \hat{g}(\xi, z_0)$$

the point $Z=\alpha$ maps to a point $\xi=\beta$ on the unit circle γ and

$$\hat{f}' = \frac{d}{d\xi} \hat{f}(\xi, z_0).$$

In the physical problem under consideration, the contour ∂B_1 must be traction free. Thus, new functions $\phi^*(z, z_0)$ and $\psi^*(z, z_0)$ must be obtained such that the tractions due to

$$\phi(z, z_0) = \hat{\phi}(z, z_0) + \phi^*(z, z_0)$$

and

$$\psi(z, z_0) = \hat{\psi}(z, z_0) + \psi^*(z, z_0)$$

vanish on the contour ∂B_1. The problem, therefore, reduces to the determination of ϕ^* and ψ^* such that

$$f^*(\beta,z_0)+\frac{w(\beta)}{\overline{w'(\beta)}}\overline{f^{*\prime}(\beta,z_0)}+\overline{g^*(\beta,z_0)}=F(\beta,z_0) \tag{10.9}$$

where, as before, $\phi^*(z,z_0)=f^*(\xi,z_0)$ and similarly for g^*.

This problem has the solution[1]

$$f^*(\xi,z_0)=\frac{1}{2\pi i}\int_\gamma \frac{F(\beta,z_0)}{\beta-\xi}d\beta - \frac{1}{2\pi i}\int_\gamma \frac{w(\beta)}{\overline{w'(\beta)}}\frac{\overline{f^{*\prime}(\beta,z_0)}}{\beta-\xi}d\beta \tag{10.10}$$

$$g^*(\xi,z_0)=\frac{1}{2\pi i}\int_\gamma \frac{\overline{F(\beta,z_0)}d\beta}{\beta-\xi} - \frac{1}{2\pi i}\int_\gamma \frac{\overline{w(\beta)}}{w'(\beta)}\frac{f^{*\prime}(\beta,z_0)}{\beta-\xi}d\beta \tag{10.11}$$

Kernels for Circular Cutout

If the contour ∂B_1 is a unit circle in the physical (z) plane, the appropriate mapping function is

$$w(\xi)=1/\xi$$

In this case, for the functions $\hat{\phi}_1$ and $\hat{\chi}_1$ of eqn. (10.5),

$$F(\beta,z_0)=-\left(\frac{1}{\beta}-z_0\right)\ln\left(\frac{1}{\beta}-z_0\right)-\frac{1}{\beta}\ln(\beta-\overline{z_0})-\frac{1}{\beta} + z_0\ln(\beta-\overline{z_0})+z_0 \tag{10.12}$$

The second term in eqn. (10.10) vanishes and the second term in eqn. (10.11) becomes $\xi^3 f^{*\prime}(\xi,z_0)$. Solving for f^* and g^*, the kernels, within additive functions of z_0, are

$$\phi_1(z,z_0)=(z-z_0)\ln(z-z_0)+z_0\ln\left(\frac{1}{z}-\bar{z}_0\right)-z\ln\left(1-\frac{1}{z\bar{z}_0}\right) \tag{10.13}$$

$$\psi_1(z,z_0)=-\bar{z}_0\ln(z-z_0)-\left(\frac{1}{z}-\bar{z}_0\right)\ln\left[\frac{1}{z}-\bar{z}_0\right]-\frac{1}{z}(1+\ln(-z_0)) + \frac{1}{z}\ln\left(1-\frac{1}{z\bar{z}_0}\right)+\frac{z_0}{z^2(1-z\bar{z}_0)}+\frac{1}{z(z\bar{z}_0-1)} \tag{10.14}$$

and $$\phi_2 = \frac{\partial \phi_1}{\partial n_0}, \ \psi_2 = \frac{\partial \psi_1}{\partial n_0}$$

Since stresses involve derivatives of these functions with respect to z, the additive functions of z_0 are of no consequence. It can be easily verified that

$$\phi_i(\alpha, z_0) + \alpha\overline{\phi_i'(\alpha, z_0)} + \overline{\psi_i(\alpha, z_0)} = 0 \qquad (i = 1,2)$$

on any point $Z = \alpha$ on the circle $|Z| = 1$, i.e. the tractions due to these functions vanish on ∂B_1.

Kernels for Elliptical Cutout

The mapping function for the case of an elliptical cutout is

$$z = w(\xi) = \frac{1}{\xi} + m\xi \tag{10.15}$$

where $m = (a-b)/(a+b)$ and $a+b=2$, in terms of the semimajor and semiminor axes, a and b, of the ellipse. This is a special case of eqn. (9.8) ($m_1 = m, m_2 = 1$) and can be easily generalized to the case $m_2 \neq 1$ (see Fig. 10.2). The functions ϕ_i and ψ_i, within additive functions of z_0, are

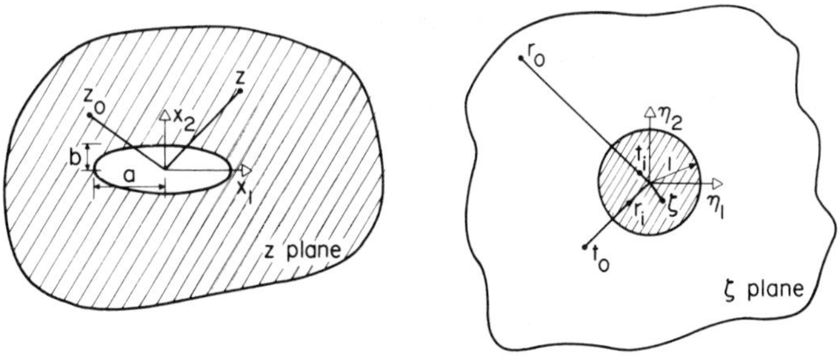

FIG. 10.2. Plate with elliptical cutout in the z plane and in the ξ plane.

$$\phi_1(z, z_0) = (z - z_0)\ln(z - z_0) - m\xi - m\xi\ln m - (m\xi - z_0)\{\ln(\xi - r_0) + \ln(\xi - t_0)\}$$
$$-\frac{1}{\xi}\{\ln(1 - \xi/r_0) + \ln(1 - \xi/t_0)\} \tag{10.16}$$
$$\psi_1(z, z_0) = -\bar{z}_0\ln(z - z_0) - \xi - \xi\ln m - (\xi - \bar{z}_0)\{\ln(\xi - r_0) + \ln(\xi - t_0)\}$$
$$-\frac{m}{\xi}\{\ln(1 - \xi/r_0) + \ln(1 - \xi/t_0)\}$$

$$+\frac{\xi^2+m}{1-m\xi^2}\left[-m\xi\{\ln m+1+\ln(\xi-r_0)+\ln(\xi-t_0)\}\right.$$

$$-\xi(\mathrm{m}\xi-z_0)\left(\frac{1}{\xi-r_0}+\frac{1}{\xi-t_0}\right)+\frac{1}{\xi}\{\ln(1-\xi/r_0)+\ln(1-\xi/t_0)\}$$

$$\left.-\frac{1}{\xi-r_0}-\frac{1}{\xi-t_0}\right] \tag{10.17}$$

$$\phi_2=\frac{\partial\phi_1}{\partial n_0},\ \psi_2=\frac{\partial\psi_1}{\partial n_0} \tag{10.18}$$

where

$$\xi=\frac{z\pm\sqrt{z^2-4m}}{2m}$$

$$r_{0,i}=\frac{z_0\pm\sqrt{z_0^2-4m}}{2m}\ \text{are the roots of}\ m\xi^2-z_0\xi+1=0$$

$$t_{0,i}=\frac{\bar{z}_0\pm\sqrt{\bar{z}_0^2-4m}}{2}\ \text{are the roots of}\ \xi^2-\bar{z}_0\xi+m=0$$

with $$|\xi|\leq 1,\ |r_0|\geq 1,\ |t_0|\geq 1.$$

10.2.3 Single Valued Displacements on Cutout Boundary

The displacement vector **u** must be single valued on the boundary of the cutout ∂B_1 in Fig. 10.3, i.e. it is required that

$$\int_{\partial B_1} \mathbf{du}=0 \tag{10.19}$$

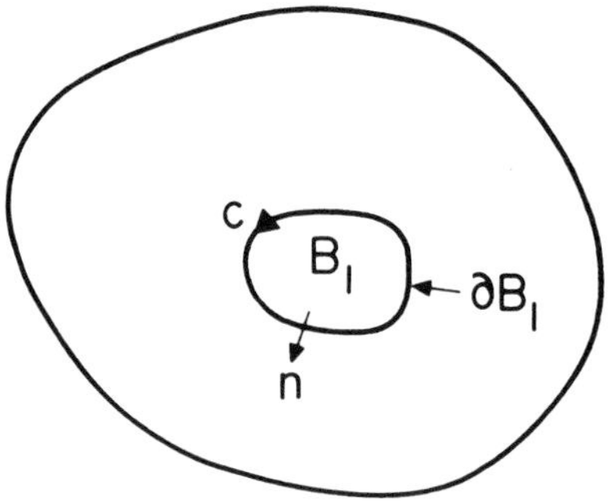

FIG. 10.3.

In the presence of nonelastic strain rates in the body but with zero tractions on ∂B_1, this condition, in terms of the rate of the stress function, gives rise to three equations for plane strain problems.

$$\int_{\partial B_1} \frac{d}{dn}(\nabla^2 \dot{\Phi})dc = \int_{\partial B_1} \mathbf{D}^{(n)} \cdot \mathbf{n} dc \tag{10.20}$$

$$\int_{\partial B_1} \left(x_2 \frac{d}{dn} - x_1 \frac{d}{dc} \right) \nabla^2 \dot{\Phi}\, dc = \int_{\partial B_1} x_2 (\mathbf{D}^{(n)} \cdot \mathbf{n}) dc - \frac{E}{1-\nu^2} \left\{ \int_{\partial B_1} \dot{\varepsilon}_{11}^{(n)} dx_1 + \dot{\varepsilon}_{12}^{(n)} dx_2 + \nu(\dot{\varepsilon}_{11}^{(n)} + \dot{\varepsilon}_{22}^{(n)})\, dx_1 \right\} \tag{10.21}$$

$$\int_{\partial B_1} \left(x_1 \frac{d}{dn} + x_2 \frac{d}{dc} \right) \nabla^2 \dot{\Phi}\, dc = \int_{\partial B_1} x_1 (\mathbf{D}^{(n)} \cdot \mathbf{n}) dc + \frac{E}{1-\nu^2} \left\{ \int_{\partial B_1} \dot{\varepsilon}_{12}^{(n)} dx_1 + \dot{\varepsilon}_{22}^{(n)} dx_2 + \nu(\dot{\varepsilon}_{11}^{(n)} + \dot{\varepsilon}_{22}^{(n)})\, dx_2 \right\} \tag{10.22}$$

where

$$D_1^{(n)} = \frac{E}{1-\nu^2} \left\{ -\frac{\partial \dot{\varepsilon}_{22}^{(n)}}{\partial x_1} + \frac{\partial \dot{\varepsilon}_{12}^{(n)}}{\partial x_2} - \nu \frac{\partial}{\partial x_1} (\dot{\varepsilon}_{11}^{(n)} + \dot{\varepsilon}_{22}^{(n)}) \right\}$$

$$D_2^{(n)} = \frac{E}{1-\nu^2} \left\{ \frac{\partial \dot{\varepsilon}_{12}^{(n)}}{\partial x_1} - \frac{\partial \dot{\varepsilon}_{11}^{(n)}}{\partial x_2} - \nu \frac{\partial}{\partial x_2} (\dot{\varepsilon}_{11}^{(n)} + \dot{\varepsilon}_{22}^{(n)}) \right\}$$

Note that $\nabla \cdot \mathbf{D}^{(n)} = C^{(n)}$.

The equations for plane stress have exactly the same form with ν set equal to zero.

These equations are derived in a manner analogous to the elastic case. The first of these equations is a statement of zero net rotation around the boundary ∂B_1, while the second and third guarantee, respectively,

$$\int_{\partial B_1} d(u_1 + x_2 \omega_{12}) = 0 \text{ and } \int_{\partial B_1} d(u_2 - x_1 \omega_{12}) = 0$$

where $\omega_{12} = u_{2,1} - u_{1,2}$ is the rotation in the plane of the body. For a discussion of the elastic situation, see, for example, Timoshenko and Goodier.[4]

It is noted that if Fig. 10.3, in fact, represented a simply connected body, the field equation (10.3) would be valid everywhere including the region B_1 and the eqns. (10.20)–(10.22) for single valued displacements would be satisfied on the boundary ∂B_1. In fact, in such a case, eqns. (10.20)–(10.22) on ∂B_1 can be derived directly from eqn. (10.3) in B_1 by using Green's theorem and the divergence theorem, provided these theorems are applicable. This matter will be alluded to in the next section.

10.2.4 Integral Equations for Stresses and Tractions

The stress rates are obtained by differentiation of eqn. (10.5) at a source point and using eqn. (10.1). For a body with a cutout (Fig. 9.3), these are (with $i, j=1, 2$)

$$\begin{aligned}8\pi\dot{\sigma}_{ij}(p)=&\int_{\partial B_2} H_{ij}^{(1)}(p,Q)C_1(Q)\,\mathrm{d}c_Q+\int_{\partial B_2} H_{ij}^{(2)}(p,Q)C_2(Q)\mathrm{d}c_Q\\&+\int_B H_{ij}^{(1)}(p,q)C^{(n)}(q)\,\mathrm{d}A_q-\int_{\partial B_1} H_{ij}^{(1)}(p,Q)D_k^{(n)}(Q)n_k(Q)\,\mathrm{d}c_Q\end{aligned} \tag{10.23}$$

where the augmented kernels $H_{ij}^{(k)}$, $k=1,2$, are

$$\begin{aligned}H_{11}^{(k)}(z,z_0)&=\mathrm{Re}\{2\phi_k'(z,z_0)-\bar{z}\phi_k''(z,z_0)-\psi_k'(z,z_0)\}\\H_{22}^{(k)}(z,z_0)&=\mathrm{Re}\{2\phi_k'(z,z_0)+\bar{z}\phi_k''(z,z_0)+\psi_k'(z,z_0)\}\\H_{12}^{(k)}(z,z_0)&=\mathrm{Im}\{\bar{z}\phi_k''(z,z_0)+\psi_k'(z,z_0)\}\end{aligned}$$

where a prime denotes differentiation with respect to z.

The functions ϕ_k and ψ_k must, of course, be chosen appropriately depending on the shape of the cutout.

The first three terms on the right-hand side are obtained from eqn. (10.5). The last term represents a layer of concentration $\mathbf{n}\cdot\mathbf{D}^{(n)}$ on the cutout boundary ∂B_1 and is included with a view towards obtaining single-valued displacements on this boundary. This term is motivated as follows. As stated in the last section, a simply connected region would require a concentration distribution $C^{(n)}$ throughout the body $B+\hat{B}$ (Fig. 9.3). Thus, if the divergence theorem is applicable in the cutout region $\hat{B}$,

$$\int_{\hat{B}} C^{(n)}\mathrm{d}A=\int_{\hat{B}} \nabla\cdot\mathbf{D}^{(n)}\mathrm{d}A=-\int_{\partial B_1} \mathbf{n}\cdot\mathbf{D}^{(n)}\mathrm{d}c$$

the negative sign being a consequence of the direction of the normal to ∂B_1 in Fig. 9.3. In the body with a cutout, however, $\hat{B}$ is a forbidden zone and $\mathbf{n}\cdot\mathbf{D}^{n}$ is distributed on ∂B_1 instead. It is postulated that inclusion of this term in eqn. (10.23) leads to satisfaction of eqn. (10.19). While a direct proof of this conjecture has not yet been possible, correct expressions for stress rates are obtained in an analytical example of a circular disc with a circular cutout, presented later in this chapter, and numerical results for a square plate with an elliptical cutout, presented in Section 10.5, agree well with those obtained from a direct formulation of the problem with Kelvin kernels of Navier's equations.[5] The corresponding elastic problem has $C^{(n)}=0$, the last two terms of eqn. (10.23) vanish, and the first two give the correct solution.

Using $C^{(n)}(q)=D^{(n)}_{k,k}(q)$ in the area integral in eqn. (10.23) and applying the divergence theorem, eqn. (10.23) can be written in a more convenient form where $\mathbf{D}^{(n)}$, with first derivatives of the strain rates, rather than $C^{(n)}$, with second derivatives, appear

$$\begin{aligned} 8\pi\dot{\sigma}_{ij}(p) = & \int_{\partial B_2} H^{(1)}_{ij}(p,Q)C_1(Q)\mathrm{d}c_Q + \int_{\partial B_2} H^{(2)}_{ij}(p,Q)C_2(Q)\mathrm{d}c_Q \\ & + \int_{\partial B_2} H^{(1)}_{ij}(p,Q)D^{(n)}_k(Q)n_k(Q)\mathrm{d}c_Q \\ & - \int_B H^{(1)}_{ij,k_0}(p,q)D^{(n)}_{k_0}(q)\cdot\mathrm{d}A_q \end{aligned} \tag{10.24}$$

Here the k_0 in the last term denotes differentiation of $H^{(1)}_{ij}$ with respect to the field point.

It is much more convenient, numerically, to have the first rather than second derivatives of nonelastic strain rates appear in the domain integral. Care must be taken, however, to see that this step does not render the kernel, $H^{(1)}_{ij,k_0}$ in this case, so strongly singular as to make the resulting calculations inaccurate or even impossible. Equation (10.24) is fine here since the starting kernel in eqn. (10.4) is of the type $r^2 \ln r$ so that $H^{(1)}_{ij,k_0}$ has a singularity of the type $1/r$.

The boundary conditions of the problem must be specified in terms of traction histories on ∂B_2. The traction rates $\dot{\tau}_i$ are obtained from eqn. (10.24) by taking the limit as p in B approaches a point P on ∂B_2. If ∂B_2 is locally smooth at P,

$$\begin{aligned} 8\pi\dot{\tau}_i(P) = & \int_{\partial B_2} H^{(1)}_{ij}(P,Q)n_j(P)C_1(Q)\mathrm{d}c_Q \\ & + \int_{\partial B_2} H^{(2)}_{ij}(P,Q)n_j(P)C_2(Q)\mathrm{d}c_Q \\ & + \int_{\partial B_2} H^{(1)}_{ij}(P,Q)D^{(n)}_k(Q)n_j(P)n_k(Q)\mathrm{d}c_Q \\ & - \int_B H^{(1)}_{ij,k_0}(P,q)D^{(n)}_{k_0}(q)n_j(P)\mathrm{d}A_q \quad (i,j,k=1,2) \end{aligned} \tag{10.25}$$

The first three integrals in the above equations must be interpreted in the sense of Cauchy principal values. It can be shown that the limiting process does not yield residues in the above equation for traction rates. In case of boundary stress rates, however, while the equations for normal and shearing stress rates do not yield a residue, the one for the tangential stress rate yields a residue of $4\pi C_2$ as p approaches P on the boundary where it is locally smooth, i.e., if $8\pi\dot{\sigma}_{cc}(p^*)=h(p^*)$ then

$$8\pi\dot{\sigma}_{cc}(P^*)=h(P^*)+4\pi C_2(P^*)$$

where p^* is infinitesimally close to P^*.

10.3 ILLUSTRATIVE EXAMPLES FOR CIRCULAR GEOMETRIES

10.3.1 An Elastic Problem for a Solid Disc (Plane Stress or Strain)[6]

The problem under consideration here is that of a solid circular disc of radius b under an axisymmetric external load p_o per unit area (Fig. 10.4). Since the region is simply connected, the formulation presented in eqn. (10.5) is used with $C^{(n)}=0$ for the elastic case. Using polar coordinates, the stresses at an inside point p are

$$8\pi\sigma_{ij}(R,\alpha)=C_1(b)\int_0^{2\pi} H_{ij}^{(1)}(R,\alpha;b,\theta)b\mathrm{d}\theta$$

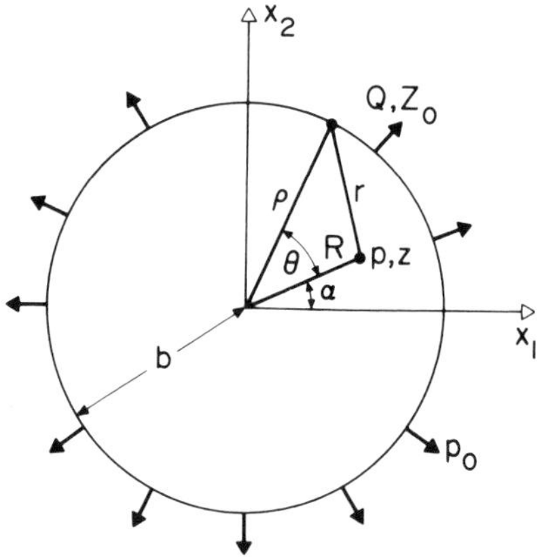

FIG. 10.4. Geometry of the elastic problem for a solid disc.

$$+C_2(b)\int_0^{2\pi} H_{ij}^{(2)}(R,\alpha;b,\theta)b\mathrm{d}\theta \tag{10.26}$$

where the kernels corresponding to K_1 and K_2 of eqn. (10.5) are

$$H_{\substack{11\\22}}^{(1)} = 2(1+\ln r) \mp \frac{R^2\cos 2\alpha + \rho^2\cos 2(\theta+\alpha) - 2\rho R\cos(\theta+2\alpha)}{r^2}$$

$$H_{12}^{(1)} = \frac{-R^2\sin 2\alpha - \rho^2\sin 2(\theta+\alpha) + 2\rho R\sin(\theta+2\alpha)}{r^2}$$

$$H_{\substack{11\\22}}^{(2)} = \frac{2(\rho - R\cos\theta)}{r^2} \pm \frac{-2\rho\cos 2(\theta+\alpha) + 2R\cos(\theta+2\alpha)}{r^2}$$

$$\pm\frac{2}{r^4}\{\rho R^2\cos 2\alpha + \rho^3\cos 2(\theta+\alpha)$$

$$-2\rho^2 R\cos(\theta+2\alpha) - R^3\cos 2\alpha\cos\theta$$

$$-\rho^2 R\cos 2(\theta+\alpha)\cos\theta + 2\rho R^2\cos(\theta+2\alpha)\cos\theta\}$$

$$H_{12}^{(2)} = \frac{-2\rho\sin 2(\theta+\alpha) + 2R\sin(\theta+2\alpha)}{r^2}$$

$$+\frac{2}{r^4}\{\rho R^2\sin 2\alpha + \rho^3\sin 2(\theta+\alpha) - 2\rho^2 R\sin(\theta+2\alpha)$$

$$-R^3\sin 2\alpha\cos\theta$$

$$-\rho^2 R\sin 2(\theta+\alpha)\cos\theta + 2\rho R^2\sin(\theta+2\alpha)\cos\theta\}$$

The symbols are shown in Fig. 10.4 and $r^2 = \rho^2 + R^2 - 2\rho R\cos\theta$. For a point P on the boundary,

$$\begin{aligned}8\pi p_o\cos\alpha &= 8\pi\tau_1(b,\alpha)\\ &= C_1(b)\int_0^{2\pi}\{H_{11}^{(1)}(b,\alpha;b,\theta)\cos\alpha + H_{12}^{(1)}(b,\alpha;b,\theta)\sin\alpha\}\,b\mathrm{d}\theta\\ &+ C_2(b)\int_0^{2\pi}\{H_{11}^{(2)}(b,\alpha;b,\theta)\cos\alpha + H_{12}^{(2)}(b,\alpha;b,\theta)\sin\alpha\}\,b\mathrm{d}\theta\end{aligned} \tag{10.27}$$

$$\begin{aligned}8\pi p_o\sin\alpha &= 8\pi\tau_2(b,\alpha)\\ &= C_1(b)\int_0^{2\pi}\{H_{12}^{(1)}(b,\alpha;b,\theta)\cos\alpha + H_{22}^{(1)}(b,\alpha;b,\theta)\sin\alpha\}\,b\mathrm{d}\theta\end{aligned}$$

$$+C_2(b)\int_0^{2\pi}\{H_{12}^{(2)}(b,\alpha;b,\theta)\cos\alpha+H_{22}^{(2)}(b,\alpha;b,\theta)\sin\alpha\}b\mathrm{d}\theta \tag{10.28}$$

The nonvanishing integrals of these kernels, used in the equations, are given in Table 10.1. Using these, both eqns. (10.27) and (10.28) give

$$2p_o=C_1b(1+\ln b)+C_2 \tag{10.29}$$

TABLE 10.1
NON-VANISHING INTEGRALS OF KERNELS FOR SOLID CIRCULAR DISC. (REFERENCE 12 HAS BEEN USED TO EVALUATE THESE INTEGRALS)

Sign		$f(\theta)$	$\int_0^{2\pi} f(\theta)\,\mathrm{d}\theta$	
$H_{11}^{(1)}$	$H_{22}^{(1)}$		$\rho>R$	$\rho=R$
+	+	$2(1+\ln r)$	$4\pi(1+\ln\rho)$	$4\pi(1+\ln R)$
$H_{11}^{(2)}$	$H_{22}^{(2)}$			
+	+	$\dfrac{2(\rho-R\cos\theta)}{r^2}$	$\dfrac{4\pi}{\rho}$	$\dfrac{2\pi}{R}$
+	−	$\dfrac{-2\rho\cos 2(\theta+\alpha)+2R\cos(\theta+2\alpha)}{r^2}$	0	$\dfrac{2\pi}{R}\cos 2\alpha$
$H_{12}^{(2)}$		$\dfrac{-2\rho\sin 2(\theta+\alpha)+2R\sin(\theta+2\alpha)}{r^2}$	0	$\dfrac{2\pi}{R}\sin 2\alpha$

and, from eqn. (10.26)

$$\sigma_{11}(R,\alpha)=\tfrac{1}{2}\{C_1b(1+\ln b)+C_2\}=p_o$$

$$\sigma_{22}(R,\alpha)=p_o,\ \sigma_{12}(R,\alpha)=0.$$

The stresses on the boundary can be obtained from eqn. (10.26) by taking the limit $p\to P$. In this case, the appropriate residues $4\pi\sin^2\alpha C_2$, $4\pi\cos^2\alpha C_2$ and $-4\pi\sin\alpha\cos\alpha C_2$ (corresponding to $4\pi C_2$ for the tangential stress $\sigma_{\theta\theta}$) must be included. This gives, for example,

$$8\pi\sigma_{11}(b,\alpha)=C_1 4\pi b(1+\ln b)b+C_2 2\pi(1+\cos 2\alpha)+C_2 4\pi\sin^2\alpha$$

and finally

$$\sigma_{11}(b,\alpha)=\sigma_{22}(b,\alpha)=p_o,\ \sigma_{12}(b,\alpha)=0$$

as expected.

10.3.2 An Inelastic Problem for a Hollow Disc (Plane Stress)[6]

A circular disc of radius b with a concentric circular cutout of unit radius is subjected to an axisymmetric external load history $p_o(t)$ (Fig. 10.5). The governing equation in polar coordinates for this problem is

$$\nabla^4\dot{\Phi}=C^{(n)}=\frac{E}{R}\frac{d}{dR}\left(\dot{\varepsilon}_{RR}^{(n)}-\dot{\varepsilon}_{\theta\theta}^{(n)}-R\frac{d\dot{\varepsilon}_{\theta\theta}^{(n)}}{dR}\right) \tag{10.30}$$

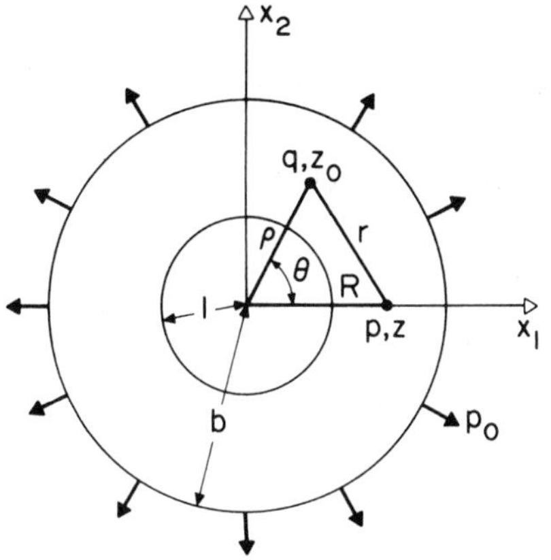

FIG. 10.5. Geometry of the inelastic problem for a hollow disc.

and

$$D_R^{(n)}=\frac{E}{R}\left(\dot{\varepsilon}_{RR}^{(n)}-\dot{\varepsilon}_{\theta\theta}^{(n)}-R\frac{d\dot{\varepsilon}_{\theta\theta}^{(n)}}{dR}\right),\quad D_\theta^{(n)}=0 \tag{10.31}$$

The equation for stress rates (10.23), with the source point p on the x_1 axis (this can be done without loss of generality for this axisymmetric problem) and $C_2(b)=0$ (using only $C_1(b)$ is sufficient here because of axisymmetry) give

$$\begin{aligned}8\pi\dot{\sigma}_{ij}(R)={}&C_1(b)\int_0^{2\pi}H_{ij}^{(1)}(R;b,\theta)b\,d\theta\\&+\int_1^b\int_0^{2\pi}H_{ij}^{(1)}(R;\rho,\theta)C^{(n)}(\rho)\rho\,d\theta\,d\rho\\&+D_R^{(n)}(1)\int_0^{2\pi}H_{ij}^{(1)}(R;1,\theta)d\theta\end{aligned} \tag{10.32}$$

For this point on the x_1 axis, $\sigma_{11}=\sigma_{RR}$, $\sigma_{22}=\sigma_{\theta\theta}$ and $\sigma_{12}=\sigma_{R\theta}$. The augmented kernels H_{ij} are obtained from eqn. (10.23) using the stress functions ϕ_1 and ψ_1 from eqns. (10.13) and (10.14). These are

$$
\begin{aligned}
H^{(1)}_{\substack{11\\22}} &= 2(1+\ln r)-2(\rho/R)\left[\frac{\cos\theta-\rho R\cos 2\theta}{r_1^2}\right]-2\{\ln r_1-\ln(\rho R)\}\\
&+\frac{2(1-\rho R\cos\theta)}{r_1^2}\mp\frac{R(R-\rho\cos\theta)}{r^2}\mp\frac{\rho}{Rr_1^4}\{-2\rho R+(1+4\rho^2R^2)\cos\theta\\
&-2\rho R(1+\rho^2R^2)\cos 2\theta+\rho^2R^2\cos 3\theta\}\mp\frac{(1-\rho R\cos\theta)}{r_1^2}\\
&\mp\frac{\rho R\cos\theta-2\rho^2R^2+\rho^3R^3\cos\theta}{r_1^4}\mp\frac{(\rho^2\cos 2\theta-\rho R\cos\theta)}{r^2}\\
&\mp\frac{2(1+\ln\rho)}{R^2}\mp\frac{(\rho R\cos\theta-1)}{R^2r_1^2}\pm\frac{\rho}{R^3r_1^4}\{-3\rho R+(2+6\rho^2R^2)\cos\theta\\
&-(4\rho R+3\rho^3R^3)\cos 2\theta+2\rho^2R^2\cos 3\theta\}\\
&\pm\frac{1}{R^2r_1^4}\{-1-4\rho^2R^2+(4\rho R+2\rho^3R^3)\cos\theta-\rho^2R^2\cos 2\theta\}
\end{aligned}
\tag{10.33}
$$

$$
\begin{aligned}
H^{(1)}_{12} &= \frac{\rho R\sin\theta}{r^2}+\frac{\rho}{Rr_1^4}\{(1+4\rho^2R^2)\sin\theta-2\rho R(1+\rho^2R^2)\sin 2\theta\\
&+\rho^2R^2\sin 3\theta\}\\
&-\frac{\rho R\sin\theta}{r_1^2}+\frac{\rho R\sin\theta(\rho^2R^2-1)}{r_1^4}+\frac{(\rho R\sin\theta-\rho^2\sin 2\theta)}{r^2}\\
&+\frac{\rho R\sin\theta}{R^2r_1^2}-\frac{\rho}{R^3r_1^4}\{(2+6\rho^2R^2)\sin\theta-(4\rho R+3\rho^3R^3)\sin 2\theta\\
&+2\rho^2R^2\sin 3\theta\}\\
&-\frac{1}{R^2r_1^4}\{2\rho^3R^3\sin\theta-\rho^2R^2\sin 2\theta\}
\end{aligned}
\tag{10.34}
$$

where

$$r^2=\rho^2+R^2-2\rho R\cos\theta$$

and

$$r_1^2=1+\rho^2R^2-2\rho R\cos\theta$$

The traction equations for a boundary point P are

$$8\pi\dot{p}_o = 8\pi\dot{t}_1(b) = C_1(b)\int_0^{2\pi} H_{11}^{(1)}(b;b,\theta)b\mathrm{d}\theta$$
$$+\int_1^b\int_0^{2\pi} H_{11}^{(1)}(b;\rho,\theta)C^{(n)}(\rho)\rho\mathrm{d}\theta\mathrm{d}\rho$$
$$+D_r^{(n)}(1)\int_0^{2\pi} H_{11}^{(1)}(b;1,\theta)\mathrm{d}\theta \qquad (10.35)$$

and the expression for $0=8\pi\dot{t}_2\ (b)$ is similar with $H_{11}^{(1)}$ replaced by $H_{12}^{(1)}$ everywhere.

The nonvanishing integrals of the kernels, in this case, are listed in Table 10.2. Using these, eqn. (10.32) on the boundary can be solved for $C_1(b)$ to give

$$C_1(b)=\frac{2b\dot{p}_o}{(b^2-1)(1+\ln b)}-D_r^{(n)}(b)-\frac{E\dot{\varepsilon}_{\theta\theta}^{(n)}(b)}{b(1+\ln b)}$$
$$-\frac{E}{b(b^2-1)(1+\ln b)}\int_1^b\left(\frac{\dot{\varepsilon}_{rr}^{(n)}-\dot{\varepsilon}_{\theta\theta}^{(n)}}{\rho}+\rho\dot{\varepsilon}_{zz}^{(n)}\right)\mathrm{d}\rho \qquad (10.36)$$

Substituting for $C_1(b)$ into eqn. (10.36) gives the equations for stress

TABLE 10.2
NON-VANISHING INTEGRALS OF KERNELS FOR ANNULAR DISC. (REFERENCE 12 HAS BEEN USED TO EVALUATE THESE INTEGRALS)

Sign		$f(\theta)$	$\int_0^{2\pi} f(\theta)\mathrm{d}\theta$		
$H_{11}^{(1)}$	$H_{22}^{(1)}$		$\rho>R$	$\rho=R$	$\rho<R$
+	+	$2(1+\ln r)$	$4\pi(1+\ln\rho)$	$4\pi(1+\ln R)$	$4\pi(1+\ln R)$
−	+	$\dfrac{R(R-\rho\cos\theta)}{r^2}$	0	π	2π
−	+	$\dfrac{\rho^2\cos 2\theta-\rho R\cos\theta}{r^2}$	0	$-\pi$	$-2\pi\rho^2/R^2$
−	+	$\dfrac{2(1+\ln\rho)}{R^2}$	$\dfrac{4\pi(1+\ln\rho)}{R^2}$	$\dfrac{4\pi(1+\ln R)}{R^2}$	$\dfrac{4\pi(1+\ln\rho)}{R^2}$

rates

$$\dot{\sigma}_{RR}(R)=\frac{E}{2}\left\{\int_1^R \frac{\dot{\varepsilon}_{RR}^{(n)}-\dot{\varepsilon}_{\theta\theta}^{(n)}}{\rho}\mathrm{d}\rho-\frac{(R^2-1)}{(b^2-1)}\frac{b^2}{R^2}\int_1^b \frac{\dot{\varepsilon}_{RR}^{(n)}-\dot{\varepsilon}_{\theta\theta}^{(n)}}{\rho}\mathrm{d}\rho\right\}$$
$$+\frac{\mathrm{E}}{2R^2}\left\{\int_1^R \rho\dot{\varepsilon}_{zz}^{(n)}\mathrm{d}\rho-\frac{(R^2-1)}{(b^2-1)}\int_1^b \rho\dot{\varepsilon}_{zz}^{(n)}\mathrm{d}\rho\right\}+\dot{p}_0\frac{(R^2-1)}{(b^2-1)}\frac{b^2}{R^2} \quad (10.37)$$

$$\dot{\sigma}_{\theta\theta}(R)=\frac{E}{2}\left\{\int_1^R \frac{\dot{\varepsilon}_{RR}^{(n)}-\dot{\varepsilon}_{\theta\theta}^{(n)}}{\rho}\mathrm{d}\rho-\frac{(R^2+1)}{(b^2-1)}\frac{b^2}{R^2}\int_1^b \frac{\dot{\varepsilon}_{RR}^{(n)}-\dot{\varepsilon}_{\theta\theta}^{(n)}}{\rho}\mathrm{d}\rho\right\}$$
$$-\frac{E}{2R^2}\left\{\int_1^R \rho\dot{\varepsilon}_{zz}^{(n)}\mathrm{d}\rho+\frac{(R^2+1)}{(b^2-1)}\int_1^b \rho\dot{\varepsilon}_{zz}^{(n)}\mathrm{d}\rho\right\}-E\dot{\varepsilon}_{\theta\theta}^{(n)}$$
$$+\dot{p}_0\frac{(R^2+1)}{(b^2-1)}\frac{b^2}{R^2} \quad (10.38)$$

$$\dot{\sigma}_{R\theta}=0$$

These equations can be derived directly from the method outlined in Section 6.4.2 (see also reference 7).

10.4 NUMERICAL IMPLEMENTATION

10.4.1 Discretization of Equations

The outer boundary of the planar body with a cutout, ∂B_2, (see Fig. 9.3) is divided into N_s straight boundary elements using $N_b (N_b=N_s)$ boundary nodes and the interior of the body, B, is divided into n_i triangular internal elements. A discretized version of eqn. (10.25) is

$$8\pi\dot{\tau}_i(P_M)=\sum_{N_s}\int_{\Delta c_i} H_{ij}^{(1)}(P_M,Q)n_j(P_M)C_1(Q)\mathrm{d}c_Q$$
$$+\sum_{N_s}\int_{\Delta c_i} H_{ij}^{(2)}(P_M,Q)n_j(P_M)C_2(Q)\mathrm{d}c_Q$$
$$+\sum_{N_s}\int_{\Delta c_i} H_{ij}^{(1)}(P_M,Q)D_k^{(n)}(Q)n_j(P_M)n_k(Q)\mathrm{d}c_Q$$
$$-\sum_{n_i}\int_{\Delta A_i} H_{ij,k_0}^{(1)}(P_M,q)D_k^{(n)}(q)n_j(P_M)\mathrm{d}A_q$$
$$(i,j,k=1,2,\ M=1,2,\ldots N_s) \quad (10.39)$$

where, as before, $\tau_i(P_M)$ are the traction components at the point P which coincides with node M and Δc_i and ΔA_i are boundary and internal elements respectively. In order to obtain the numerical results presented in the next section, the concentrations C_1 and C_2 are assumed to vary linearly over each boundary element with (generally) their values assigned at the nodes which lie at the intersections of these elements. Possible discontinuities in C_1 and C_2, across corners for example, are allowed for by using source (sampling) points away from the corner on boundary elements that meet at a corner. A double source point at a corner itself cannot be used with this formulation since this leads to two identical algebraic equations after discretization, and therefore, to a singular matrix.

The nonelastic strain rates $\dot{\varepsilon}_{jk}^{(n)}$ are interpolated linearly over each triangular internal element. Hence, the components of $\mathbf{D}^{(n)}$ are uniform within each element. Integrals of $H_{ij}^{(k)}$ and $cH_{ij}^{(k)}$ on boundary elements are evaluated by Gaussian integration, except analytically for singular terms. Integrals on internal elements are always evaluated by Gaussian quadrature. As in Chapter 9, this method is adequate for these problems with the source points lying on the vertices of the triangles.

Substitution of the linear functional forms of C_1 and C_2 into eqn. (10.39) leads to an algebraic system of the type

$$\{\dot{\tau}\} = [A]\{\mathbf{x}\} + \{\mathbf{d}\} \tag{10.40}$$

The coefficients of the matrix $[A]$ contain boundary integrals of the kernels. The traction rates are prescribed, the vector $\{\mathbf{d}\}$ contains integrals of the kernels and the nonelastic strain rates and the vector $\{\mathbf{x}\}$ contains the unknown values of the concentrations at the boundary nodes.

Equation (10.24) for stress rates at an internal point p is discretized in a similar fashion.

The initial values of the state variables are prescribed and the initial stress field is obtained by solving the corresponding elastic problem. The initial rates of the nonelastic strain rates and state variables are obtained from a suitable constitutive model. The vector $\mathbf{d}$ in eqn. (10.40) is calculated next, and this, together with the prescribed rates of boundary tractions, are used to calculate the initial values of the boundary concentrations C_1 and C_2. These concentrations are now used in a discretized version of eqn. (10.24) to calculate the initial stress rates throughout the body. These rates are used to find the values of the variables after a small time interval Δt, and so on, and in this way the time histories of the relevant variables are obtained. Once again, time integration is carried out by an Euler type time-integration strategy with automatic time-step control, as discussed in Section 4.2.1.

10.5 NUMERICAL RESULTS AND DISCUSSION

10.5.1 Constitutive Model and Loading

In all but one example (Fig. 10.6), an elastic-power-law creep constitutive model is used to describe material behavior. The material parameters for

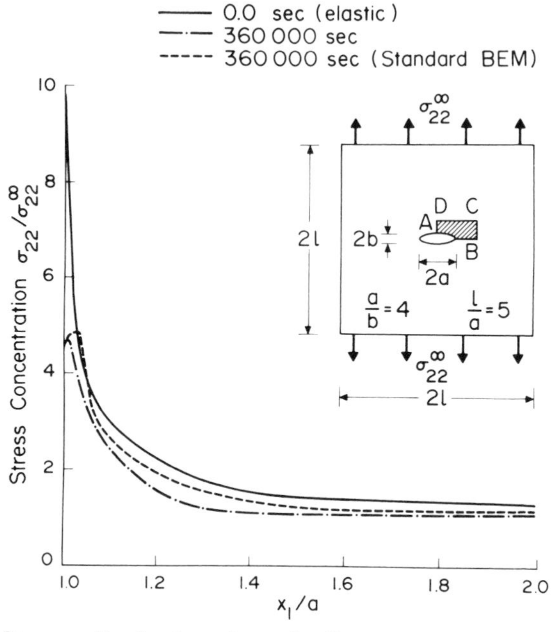

FIG. 10.6. Stress redistribution along the line $x_2=0$ in plate with elliptical cutout of axis ratio 4:1. Uniaxial tension $\sigma_{22}^{\infty}=4$ ksi. Hart's model.

this model, for stainless steel at 400° C, are the same as those listed in Section 9.6.1. The values of n used are 2, 4 and 7. The elastic properties used are

$$E=0{\cdot}244\times10^{8}\ \text{psi} \qquad \nu=0{\cdot}298$$

Two kinds of Mode I remote loading σ_{22} are considered here. The first is time-independent remote stress and the second is remote stress increasing from zero at a constant rate.

10.5.2 Computer Program Validation

The redistribution of stress in a square plate with an elliptical cutout of axis ratio 4, loaded by a remote stress of 4000 psi (Fig. 10.6) is obtained by two methods. The first (method (a)[5]) is the usual planar BEM formulation in terms of displacements as outlined in Chapter 5 and the other (method (b)) is the present formulation with augmented kernels. For the present method, quarter symmetry is used with 12 boundary nodes distributed uniformly along the edges $x_1 = l$, $0 \leq x_2 \leq l$ and $x_2 = l$, $0 \leq x_1 \leq l$ of the plate. The internal cell distribution for this problem is given in reference 8.

The results for $\sigma_{22}/\sigma_{22}^{\infty}$ as functions of x_1, at various times, are shown in Fig. 10.6. The results from methods (a) and (b), at zero time, coincide within plotting accuracy and those at 100 hours agree quite well. Method (b), which uses a piecewise linear interpolation of nonelastic strain rates, is expected to be more accurate than method (a) which uses piecewise uniform nonelastic strain rates.

10.5.3 Crack Model and Discretization

As in Chapter 9, a crack is modelled as a thin elliptical cutout in the center of a square plate. Only an axis ratio $a/b = 199$ is used here. The region very near the tip of the major axis of the ellipse is discretized into internal cells as shown in Fig. 9.4. The loading, of course, is tensile in the 2 direction as in Fig. 10.6. Quarter symmetry is used and 12 boundary nodes are distributed uniformly along the edges of the plate in the first quadrant.

10.5.4 Numerical Results for Mode 1, Plane Stress—Comparison with Asymptotic Results[9]

Numerical results for Mode 1, plane stress problems are shown in Figs. 10.7–10.11. Figure 10.7 shows the stress concentration as a function of time for a constant remote load. The other figures are for remote load increasing at a uniform rate. Also, Figs. 10.7–10.9 show the time variation of stress concentration at a crack tip while the others show the redistribution of stress near a crack tip as functions of position and time.

The results are qualitatively similar to those reported for Mode III in Chapter 9. Asymptotic results for Mode I stress variation[10–11] have the same dependence on distance ρ from the crack tip and time as for Mode III

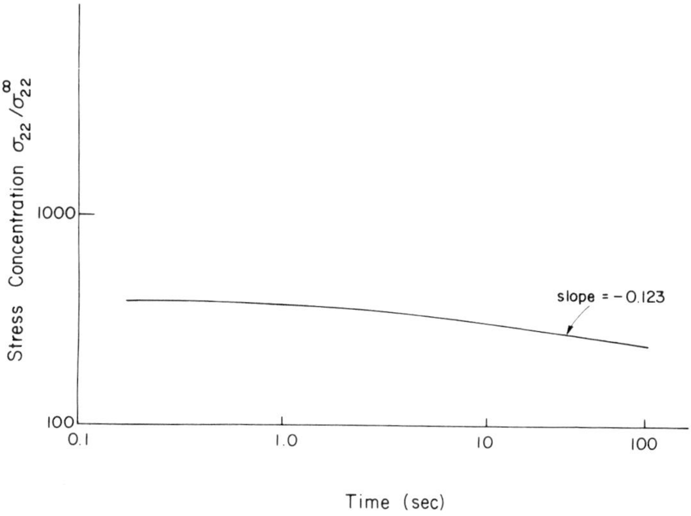

FIG. 10.7. Stress concentration at crack tip as a function of time for constant remote stress $\sigma_{22}^{\infty}=500$ psi, $n=7$, $a/b=199$.

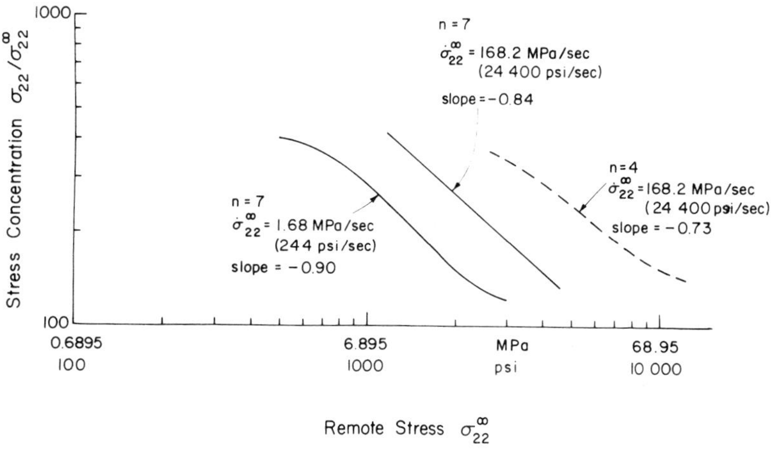

FIG. 10.8. Stress concentration at crack tip as function of remote stress for remote stress increasing at a constant rate. $a/b=199$.

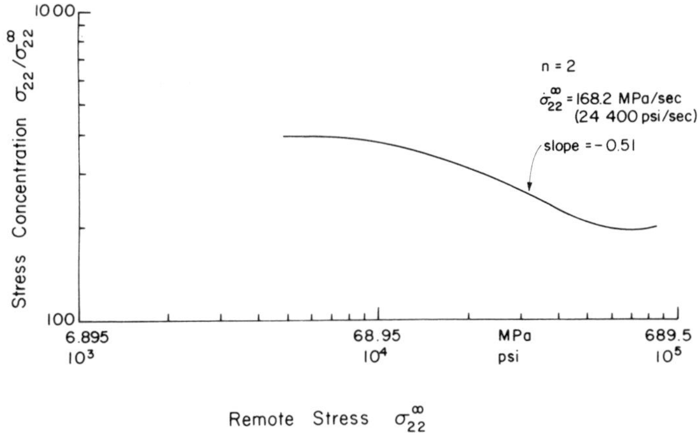

FIG. 10.9. Same situation as in Fig. 10.8. $n=2$, $a/b=199$.

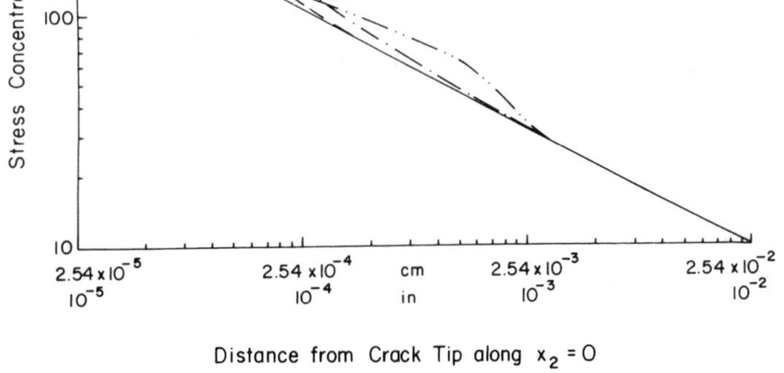

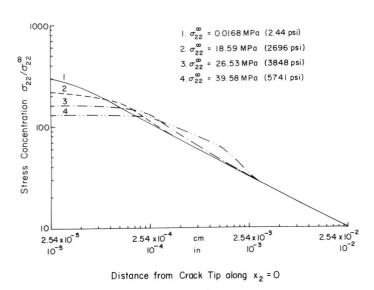

FIG. 10.10. Stress redistribution near crack tip for remote stress increasing at a constant rate. $\dot{\sigma}_{22}^{\infty} = 24\,400$ psi/sec. $n=7$, $a/b=199$.

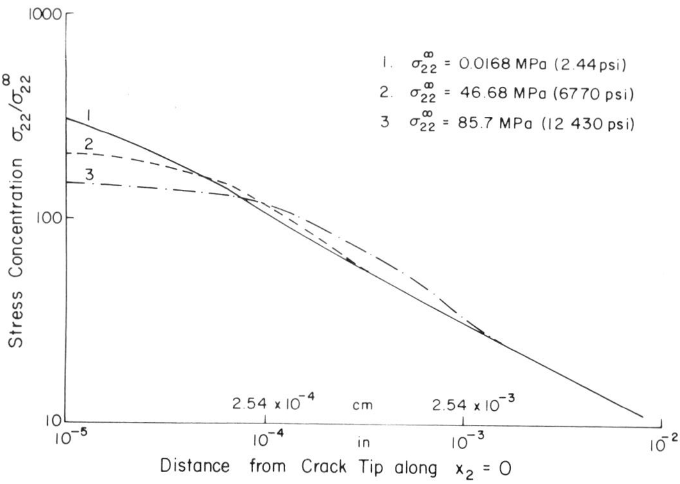

FIG. 10.11. Same situation as in Fig. 10.10 with $n=4$, $a/b=199$.

(see section 9.5.1). A comparison of measured slopes from the numerical solutions and asymptotic results is given in Table 10.3. The correlation of slopes is quite good except for the case $n=2$ (Fig. 10.9). This is not surprising in view of the fact that the asymptotic analysis neglects elastic strain rates near a crack tip relative to the nonelastic ones, and this assumption becomes weaker for small values of the creep exponent n.

The spatial variation of stress near a crack tip, as obtained from the BEM and the asymptotic calculations, cannot be compared quantitatively in view of the fact that the asymptotic results are obtained for a line crack while the BEM results are valid for a crack modelled as a very thin ellipse. While this difference in modelling does not affect the stress distribution slightly away from a crack tip, it does change the local variation at and very near a crack tip. There is, however, qualitative agreement between the results. According to the asymptotic results, the stress concentration $\sigma_{22}/\sigma_{22}^{\infty}$ should vary as $\rho^{-1/(n+1)}$ in terms of the distance ρ from a crack tip. Curve 4, in Fig. 10.10, shows that $\sigma_{22}/\sigma_{22}^{\infty}$ is substantially independent of ρ, for $\rho <$ about $0{\cdot}6 \times 10^{-4}$ in and $n=7$. A similar curve for $n=2$ (given in reference 9, Fig. 13) shows a much stronger dependence of stress concentration on ρ in the same region.

It should be mentioned, in conclusion, that the BEM analysis of inelastic fracture, as described in Chapters 9 and 10, is ideally suited for the study of the general transient problem. The method is quite general

TABLE 10.3

DEPENDENCE OF STRESS CONCENTRATION AT CRACK TIP WITH TIME FOR MODE I

Figure	*a/b*	*Loading*	*n*	*Slope (analytical)*	*Slope (numerical)*	*Slope* $\frac{(Analyt\text{-}Num)}{Analyt} \times 100$
10·7	199	$\sigma_{22}^{\infty} = 500$ psi	7	$-1/(n+1) = -0{\cdot}125$	−0·13	−4
10·8	199	$\dot{\sigma}_{22}^{\infty} = 24{\cdot}4$ ksi/sec	7	$-n/(n+1) = -0{\cdot}875$	−0·84	4
10·8	199	$\dot{\sigma}_{22}^{\infty} = 244$ psi/sec	7	−0·875	−0·90	−2·8
10·8	199	$\dot{\sigma}_{22}^{\infty} = 24{\cdot}4$ ksi/sec	4	−0·80	−0·73	8·75
10·9	199	$\dot{\sigma}_{22}^{\infty} = 24{\cdot}4$ ksi/sec	2	−0·667	−0·51	23·5

and can be used to solve problems with

(a) Plates of any size or shape.
(b) Any crack orientation and loading, i.e. mixed mode situations.
(c) General time-dependent remote loading.
(d) Other constitutive models for material deformation which fit the general mathematical structure outlined in Chapter 2.

REFERENCES

1. MUSKHELISHVILI, N. I. *Some Basic Problems of the Mathematical Theory of Elasticity*, (trans. J.R.M. Radok), Noordhoff, Groningen, Holland (1953).
2. MIR-MOHAMAD SADEGH, A. and ALTIERO, N. J. A boundary-integral approach to the problem of an elastic region weakened by an arbitrarily shaped hole. *Mechanics Research Communications*, **6**, 167–175 (1979).
3. MIR-MOHAMAD SADEGH, A. and ALTIERO, N. J. Solution of the problem of a crack in a finite region using an indirect boundary-integral method. *Engineering Fracture Mechanics*, **11**, 831–837 (1979).
4. TIMOSHENKO, S. P. and GOODIER, J. N. *Theory of Elasticity* (3rd ed.), McGraw-Hill, New York (1970).
5. MORJARIA, M. and MUKHERJEE, S. Improved boundary-integral equation method for time-dependent inelastic deformation in metals. *International Journal for Numerical Methods in Engineering*, **15**, 97–111 (1980).
6. MUKHERJEE, S. and MORJARIA, M. A boundary element formulation for planar time-dependent inelastic deformation of plates with cutouts. *International Journal of Solids and Structures*, **17**, 115–126 (1981).
7. MUKHERJEE, S. Thermoviscoplastic response of cylindrical structures using a state variable theory. In *Mechanical Behavior of Materials—Proceedings of ICM–3*, Cambridge, England. K. J. Miller and R. F. Smith (eds.), Pergamon Press, Oxford and New York, **2**, 233–242 (1979).
8. MORJARIA, M. and MUKHERJEE, S. Numerical analysis of planar, time-dependent inelastic deformation of plates with cracks by the boundary element method. *International Journal of Solids and Structures*, **17**, 127–143 (1981).
9. MORJARIA, M. and MUKHERJEE, S. Numerical solutions for stresses near crack tips in time-dependent inelastic fracture mechanics. *International Journal of Fracture*, **18**, 293–310 (1982).
10. RIEDEL, H. Cracks loaded in anti-plane shear under creep conditions, *Zeitschrift für Metallkunde*, **69**, 755–760 (1978).
11. RIEDEL, H. and RICE, J. R. Tensile cracks in creeping solids. American Society for Testing and Materials, *Special Technical Publication*, 700, 112–130 (1980).
12. GRADSHTEYN, I. S. and RYZHIK, I. W. *Table of Integrals, Series and Products.* Academic Press, New York (1965).

Index